DAILY QUIZZES

WITH ANSWER KEY

HOLT
PEOPLE, PLACES, AND CHANGE

An Introduction To World Studies

HOLT, RINEHART AND WINSTON

A Harcourt Education Company

Austin · Orlando · Chicago · New York · Toronto · London · San Diego

Printed in the United States of America

ISBN 0-03-066694-5

1 2 3 4 5 6 7 8 9   082   04 03 02

TABLE OF CONTENTS

TABLE OF CONTENTS

A Geographer's World

CHAPTER 1

SECTION 1

True/False • *(10 points each)* Indicate whether each statement below is true or false by writing *T* or *F* in the space provided. If the statement is false, use the space below to explain why.

_____ **1.** All people share the same perspective on the world.

_____ **2.** Geographers use spatial perspective to look at the world.

_____ **3.** Scientists have determined that the study of geography has no application to real life.

_____ **4.** Geographers study a wide variety of Earth's processes and their impact on people.

_____ **5.** One area of interest to geographers is how governments change and how those changes affect people and their lives.

_____ **6.** Geographers only study issues at the global level.

_____ **7.** One way for students to learn geography is for them to study their communities at the local level.

_____ **8.** Regional studies cover smaller areas than studies done at the local level.

_____ **9.** Urban areas contain cities.

_____ **10.** Rural areas contain open land that is often used for farming.

Name _____ Class _____ Date _____

A Geographer's World

Matching • *(10 points each)* Match each of the following terms with the correct description by writing the letter of the description in the space provided. Some descriptions will not be used.

_____ **1.** absolute location

_____ **2.** relative location

_____ **3.** physical features

_____ **4.** adaptations

_____ **5.** levees

_____ **6.** diffusion

_____ **7.** regions

_____ **8.** subregions

_____ **9.** human systems

_____ **10.** environment

a. Changes people make in their own behavior so that they are better suited to live in an environment

b. For example, language, religion, history, or other cultural qualities studied by geographers

c. An exact spot on Earth

d. The location of one's parents, siblings, or children

e. Smaller regions

f. Large walls usually built from dirt along rivers to help stop flooding

g. Position of a place in relation to another place

h. Movement of ideas or behaviors from one cultural region to another

i. Areas of Earth's surface with one or more shared characteristics that make them different from the surrounding areas

j. For example, landforms, coastlines, types of soils, lakes, or rivers

k. Conditions and features surrounding people or places

l. Migration of people between countries

A Geographer's World

SECTION 3

Fill in the Blank • *(10 points each)* For each of the following statements, fill in the blank with the appropriate word, phrase, or name.

1. The subject matter of _____ is people—past or present, their location and distribution over Earth, their activities, and their differences.

2. _____ geographers study the exchange of goods and services across Earth.

3. The study of Earth's natural landscapes and physical systems, including the atmosphere, is the subject matter of _____.

4. Mountains, plains, and deserts are different types of _____.

5. If you work in your hometown, you should know your _____ geography.

6. _____ is the art and science of mapmaking.

7. Today, most mapmakers do their work on _____.

8. The field of forecasting and reporting temperature, rainfall, and other atmospheric

   conditions is known as _____.

9. The field of _____ focuses on Earth's larger atmospheric systems.

10. Scientists who track and study Earth's larger atmospheric systems are known as

   _____.

CHAPTER 2 Planet Earth

Matching • *(10 points each)* Match each of the following terms with the correct description by writing the letter of the description in the space provided. Some descriptions will not be used.

_____ **1.** solar system

_____ **2.** satellite

_____ **3.** axis

_____ **4.** revolution

_____ **5.** Arctic Circle

_____ **6.** Antarctic Circle

_____ **7.** solstice

_____ **8.** atmosphere

_____ **9.** hydrosphere

_____ **10.** biosphere

a. Day when the Sun's vertical rays are farthest from the equator

b. One complete spin of Earth on its axis

c. The Sun and the objects that move around it

d. Part of the Earth system that includes all plant and animal life

e. Imaginary line running from the North Pole through the center of the planet to the South Pole

f. Part of the Earth system that consists of all of Earth's water, found in places like lakes, oceans, and glaciers

g. Line of latitude located 66.5° south of the equator and that circles the South Pole

h. Gas that helps protect Earth from harmful radiation

i. One orbit of Earth around the Sun

j. Line of latitude located 66.5° north of the equator and that circles the North Pole

k. Layer of gases—the air—that surrounds Earth

l. Body that orbits a larger body

CHAPTER **2**

Planet Earth

Fill in the Blank • *(10 points each)* For each of the following statements, fill in the blank with the appropriate word, phrase, or name.

1. _____ is the only substance on Earth that occurs naturally as a solid, a liquid, and a gas.

2. The circulation of water from Earth's surface to the atmosphere and back is known

   as the _____ cycle.

3. As water vapor rises, it cools, causing _____, the process by which water changes from a gas into tiny liquid droplets.

4. Some states in the United States have few natural lakes, so they have dammed rivers

   to create artificial lakes called _____.

5. A(n) _____ is a smaller stream or river that flows into a larger stream or river.

6. _____ is water from rainfall, rivers, lakes, and melting snow that seeps into the ground, filling in spaces between soil and grains of rock.

7. Most of Earth's water is found in the _____.

8. The deepest place in the ocean, the Mariana Trench in the _____ Ocean, is about 36,000 feet deep.

9. Heavy rains can cause _____, which are the world's deadliest natural hazard.

10. _____, like those on the Nile River and the Colorado River, are one way that people control floods.

Planet Earth

CHAPTER 2

Multiple Choice • *(10 points each)* For each of the following, write the letter of the *best* choice in the space provided.

____ **1.** Land bordered by water on three sides is called a
    a. plain.   **c.** delta.
    b. fault.   **d.** peninsula.

____ **2.** What is a plateau?
    a. an elevated flatland
    b. a neck of land connecting two larger areas of land
    c. a nearly flat area
    d. the inner, solid part of the planet

____ **3.** The outer, solid layer of Earth is the
    a. crust.
    b. plate tectonic.
    c. mantle.
    d. core.

____ **4.** Which of the following are common in subduction zones?
    a. floods
    b. earthquakes
    c. monsoons
    d. hurricanes

____ **5.** Because it is rimmed by active volcanoes, the Pacific Plate's edge has been called the
    a. Home of Heat.
    b. Land of Lava.
    c. Ring of Fire.
    d. Sea of Magma.

____ **6.** The Himalayas are
    a. Earth's two largest volcanoes.
    b. the world's highest mountain range.
    c. huge ocean waves caused by undersea earthquakes.
    d. plateaus in Australia.

____ **7.** Which of the following is similar to an underwater mountain range?
    a. a delta
    b. a continental drift
    c. a subduction zone
    d. a mid-ocean ridge

____ **8.** Pangaea is the name used for
    a. the deepest part of a river.
    b. Earth's original supercontinent.
    c. a recently discovered planet.
    d. the worst hurricane in history.

____ **9.** Weathering and erosion change primary landforms into
    a. secondary landforms.
    b. lava.
    c. mantles.
    d. continents.

____ **10.** Large, slow-moving rivers of ice are called
    a. faults.   **c.** glaciers.
    b. cores.   **d.** sediments.

Wind, Climate, and Natural Environments

Fill in the Blank • *(10 points each)* For each of the following statements, fill in the blank with the appropriate word, phrase, or name.

1. All planets in our solar system receive energy from the _____.

2. When the direct rays of the Sun strike Earth at the _____, it is summer in the Northern Hemisphere.

3. The _____ is the process by which the atmosphere traps heat.

4. When the _____ is blowing, we know that air is moving from one place to another.

5. The term _____ refers to the weight of the air.

6. An instrument called a(n) _____ is used to measure air pressure.

7. When a large amount of warm air meets a large amount of cool air, an area of unstable weather, called a(n) _____, forms.

8. Winds that blow in the same direction over large areas of Earth are called _____ winds.

9. Warm ocean water from the tropics moves in giant streams, or _____, to colder areas.

10. The Gulf Stream is a(n) _____ current.

Wind, Climate, and Natural Environments

SECTION 2

True/False • *(10 points each)* Indicate whether each statement below is true or false by writing *T* or *F* in the space provided. If the statement is false, use the space below to explain why.

_____ **1.** The condition of the atmosphere in a local area for a short period of time is called weather.

_____ **2.** The weather in a community over a long period of time is that community's climate.

_____ **3.** A humid tropical climate is cold and dry all year long.

_____ **4.** A monsoon may be wet or dry.

_____ **5.** If an area is arid, it receives a great deal of rain.

_____ **6.** The steppe climate is found between desert and wet climate regions.

_____ **7.** The southeastern United States provides an example of the Mediterranean climate.

_____ **8.** Boreal forests are found only in the Southern Hemisphere.

_____ **9.** Permafrost is found only in desert climates.

_____ **10.** The polar regions of Earth have what is known as an ice cap climate.

Wind, Climate, and Natural Environments

CHAPTER 3

SECTION 3

Matching • *(10 points each)* Match each of the following terms with the correct description by writing the letter of the description in the space provided. Some descriptions will not be used.

_____ **1.** extinct

_____ **2.** ecology

_____ **3.** photosynthesis

_____ **4.** roots

_____ **5.** food chain

_____ **6.** nutrients

_____ **7.** plant communities

_____ **8.** ecosystem

_____ **9.** plant succession

_____ **10.** humus

a. Parts of a plant that take in sunlight and carbon dioxide from the air

b. Having died out completely

c. Gradual process by which one group of plants replaces another group of plants

d. All of the plants and animals in an area together with the nonliving parts of their environment

e. Study of the connections among different forms of life

f. Term for decayed plant and animal material in soil

g. Substances that promote growth

h. Parts of a plant that take in minerals, water, and gases from the soil

i. Process in which topsoil is swept away by water or wind

j. Groups of plants that live in the same area

k. Arrangement of living things in a community according to the order in which each uses the next lowest member as a source of food

l. Process by which plants convert sunlight into chemical energy

Earth's Resources

True/False • *(10 points each)* Indicate whether each statement below is true or false by writing *T* or *F* in the space provided. If the statement is false, use the space below to explain why.

_____ **1.** Soil and forests are renewable resources, or resources that can be replaced by Earth's natural processes.

_____ **2.** Soil consists of rock particles and humus as well as water and gases.

_____ **3.** Farmers use fertilizers to kill pests that eat vegetation.

_____ **4.** Crop rotation refers to the system of growing different crops on the same land over a period of years.

_____ **5.** Having a great deal of salt in the soil helps crops to grow.

_____ **6.** Building terraces into hillsides helps stop soil from being washed away.

_____ **7.** The long-term process of losing soil fertility and plant life is desertification.

_____ **8.** Forests are important because they provide both people and wildlife with food and shelter.

_____ **9.** Since the 1890s reforestation has no longer been possible.

_____ **10.** Deforestation is occurring only in the United States.

Earth's Resources

CHAPTER 4

Fill in the Blank • *(10 points each)* For each of the following statements, fill in the blank with the appropriate word, phrase, or name.

1. _____ regions are regions that receive a small amount of rain.

2. People who live in drier regions have built _____—artificial channels for carrying water—to bring in fresh water.

3. Some places have water deep underground in _____.

4. The process of _____ involves removing salt from seawater.

5. The water in industrialized countries like the United States is being polluted when farmers use too much chemical fertilizer and _____.

6. Rivers carry pollution to the _____, where it can harm marine life.

7. Cities like Denver, Los Angeles, and Mexico City have special problems with air pollution because they are located in bowl-shaped _____ that trap air pollution.

8. When air pollution combines with moisture in the air and then falls to the ground, it creates _____, which can kill trees and even fish.

9. The _____ layer absorbs harmful radiation from the Sun.

10. The slow increase in Earth's average temperature is called _____.

Earth's Resources

SECTION 3

Matching • *(10 points each)* Match each of the following terms with the correct description by writing the letter of the description in the space provided. Some descriptions will not be used.

_____ **1.** nonrenewable resources

_____ **2.** minerals

_____ **3.** metallic minerals

_____ **4.** mercury

_____ **5.** gold

_____ **6.** iron

_____ **7.** aluminum

_____ **8.** nonmetallic minerals

_____ **9.** diamond

_____ **10.** sulfur

a. Metal that is liquid at room temperature

b. Minerals that have a dull surface and are poor conductors of heat and electricity

c. Resources that cannot be replaced by natural processes or are replaced very slowly

d. Metallic mineral used in making the penny

e. Minerals that are shiny and can conduct heat and electricity

f. The cheapest of all metals

g. Mineral substance used to make batteries

h. Substances that are part of Earth's crust

i. Mineral made of pure carbon that is the hardest naturally occurring substance known

j. Resources that can be replaced by Earth's natural processes

k. Lightweight metal used for soft drink cans

l. One of the heaviest of all metals

Name _____ Class _____ Date _____

Earth's Resources

Multiple Choice • *(10 points each)* For each of the following, write the letter of the *best* choice in the space provided.

_____ **1.** Fossil fuels are
a. metallic minerals.
b. renewable resources.
c. nonmetallic minerals.
d. nonrenewable resources.

_____ **2.** Which of the following are fossil fuels?
a. coal and natural gas
b. gold and platinum
c. sun and wind
d. water and air

_____ **3.** Sources of energy used until the 1900s mainly were
a. petroleum and wind.
b. coal and wood.
c. natural gas and water.
d. sun and electricity.

_____ **4.** Which of the following energy sources appears in the form of an oily liquid?
a. petroleum
b. coal
c. natural gas
d. none of the above

_____ **5.** Of the crude oil that we know about, some 65 percent is found in
a. South America.
b. Australia.
c. Southwest Asia.
d. Eastern Europe.

_____ **6.** The cleanest-burning fossil fuel is
a. wood.
b. petroleum.
c. natural gas.
d. coal.

_____ **7.** The most widely used renewable energy source is
a. geothermal energy.
b. solar energy.
c. wind power.
d. hydroelectric power.

_____ **8.** What percentage of the electricity used in the United States is produced by dams?
a. 9 percent
b. 25 percent
c. 50 percent
d. 75 percent

_____ **9.** Geothermal energy is produced by
a. the Sun's heat and light.
b. Earth's internal heat.
c. water.
d. wind turbines.

_____ **10.** The United States does not intend to increase its use of
a. solar energy.
b. geothermal energy.
c. hydroelectric power.
d. nuclear energy.

CHAPTER 5

The World's People

Fill in the Blank • *(10 points each)* For each of the following statements, fill in the blank with the appropriate word, phrase, or name.

1. _____ is a learned system of shared beliefs and ways of doing things that guide a person's daily behavior.

2. When a cultural group shares beliefs and practices learned from parents, grandparents, and ancestors, it is sometimes called a(n) _____.

3. When people from different cultures live in the same country, the country is said to be multiethnic or _____.

4. _____ is based on inherited physical or biological traits.

5. Shared elements of culture such as dress, food, or religious beliefs are called

   _____.

6. When people make important cultural changes as a result of long-term contact with another society, _____ occurs.

7. Two important factors that influence the way that people meet basic needs are their history and _____.

8. The process by which humans selected plants and tamed animals, making them dependent on people, is called _____.

9. _____ is farming that provides all or almost all of the food needed by a farm family with very little left over.

10. The term for a culture that has become highly complex is _____.

The World's People

SECTION 2

Matching • *(10 points each)* Match each of the following terms with the correct description by writing the letter of the description in the space provided. Some descriptions will not be used.

_____ **1.** population density

_____ **2.** traditional economy

_____ **3.** unlimited government

_____ **4.** tertiary industries

_____ **5.** quaternary industries

_____ **6.** command economy

_____ **7.** developing countries

_____ **8.** free enterprise

_____ **9.** democracy

_____ **10.** communism

a. Industries that handle goods that are ready to be sold to consumers

b. Number of births per 1,000 people in a year

c. Measured by dividing a country's population by its area

d. System of government in which the government controls industries, prices, and wages

e. Industries whose activities directly involve natural resources or raw materials

f. System of government in which voters elect leaders and rule by majority

g. Countries that are in different stages of moving toward being developed

h. Governments with total control over their citizens

i. Type of economy in which companies are free to make whatever goods they wish and to sell them at any price buyers will pay

j. Economy based on custom and tradition

k. Economy in which government decides what and how much will be produced

l. Industries in which workers have specialized skills or knowledge

The World's People

True/False • *(10 points each)* Indicate whether each statement below is true or false by writing *T* or *F* in the space provided. If the statement is false, use the space below to explain why.

_____ **1.** Carrying capacity refers to the greatest number of individuals that the resources of an area can support.

_____ **2.** Population growth rates are the same in all areas of the world.

_____ **3.** In most developing countries, populations continue to grow rapidly.

_____ **4.** In general, a high population growth rate will help a country's economic development.

_____ **5.** Natural resources such as fresh water, minerals, and fertile land are distributed evenly over Earth.

_____ **6.** Japan has few energy resources but is a world leader in manufacturing.

_____ **7.** The need for scarce resources usually leads countries to trade peacefully.

_____ **8.** Some people believe that Earth can easily support a much larger human population.

_____ **9.** The amount of land available for farming is growing.

_____ **10.** Oil is a renewable resource and new supplies will never run out.

CHAPTER 6 — The United States

Fill in the Blank • *(10 points each)* For each of the following statements, fill in the blank with the appropriate word, phrase, or name.

1. The 48 _____ American states and the District of Columbia lie between the Atlantic Ocean and the Pacific Ocean.

2. The _____ include mountain ranges and river valleys from the state of Maine to the state of Alabama.

3. The Great Lakes are made up of Lake Superior, Lake Huron, Lake Erie, Lake Michigan, and Lake _____.

4. Running along the crest of the Rockies is the _____, which divides the rivers of North America into those that flow east and those that flow west.

5. The highest mountain in North America is Mount _____ in the Alaska Range.

6. Southern Florida has a _____ climate and is warm all year.

7. The Great Plains, which has a _____ climate, supports wide grasslands.

8. With the exception of the southeast, most of the state of Alaska has very cold, subarctic and _____ climates.

9. Some of the most productive farmlands in the world can be found in the Interior _____ of the United States.

10. Alaska, California, Texas, and other states in the country are important sources of oil and natural _____.

The United States

Matching • *(10 points each)* Match each of the following terms or places with the correct description by writing the letter of the description in the space provided. Some descriptions will not be used.

_____ **1.** Anasazi

_____ **2.** Iroquois

_____ **3.** colony

_____ **4.** Britain

_____ **5.** Constitution

_____ **6.** Spanish

_____ **7.** bilingual

_____ **8.** Hanukkah

_____ **9.** Kwanzaa

_____ **10.** blues

a. Having the ability to speak two languages

b. Holiday celebrated by the Jewish people

c. Large farm that grows mainly one crop

d. People in the southwestern United States who built a complex irrigation system around A.D. 700

e. Musical form that originated in the United States

f. Territory controlled by people from a foreign land

g. Holiday celebrated by African Americans, based on a traditional African festival

h. Country that the 13 original American colonies broke away from in 1776

i. Language spoken by about 17 million U.S. residents

j. Mexican holiday celebrated by Mexican Americans

k. People in the Northeast who hunted and gathered wild foods

l. Basis of the limited, democratic government of the United States

The United States

Multiple Choice • *(10 points each)* For each of the following, write the letter of the *best* choice in the space provided.

____ 1. Which of the following regions has 12 states and shares a border with Canada?
   a. Midwest
   b. Northeast
   c. South
   d. Pacific

____ 2. The first industrial area in the United States was
   a. the Great Plains.
   b. the Deep South.
   c. the Great Northwest.
   d. New England.

____ 3. States found in the South include
   a. Arkansas, Louisiana, and Texas.
   b. Colorado, Nevada, and Wyoming.
   c. New York, Vermont, and Delaware.
   d. Maine, Montana, and Massachusetts.

____ 4. The South has long coastlines along the Atlantic Ocean and the
   a. Great Lakes.
   b. Gulf of Mexico.
   c. Mediterranean Sea.
   d. Pacific Ocean.

____ 5. The Corn Belt is found in the
   a. Interior West.   c. South.
   b. Midwest.   d. Pacific.

____ 6. States in the Dairy Belt are major producers of
   a. lumber.
   b. cotton.
   c. coal.
   d. milk.

____ 7. States in the Wheat Belt include
   a. Florida and Georgia.
   b. Washington and Oregon.
   c. Nebraska and Oklahoma.
   d. Rhode Island and Massachusetts.

____ 8. Yellowstone National Park is found in the
   a. South.
   b. Pacific.
   c. Interior West.
   d. Midwest.

____ 9. What percentage of Americans live in the state of California?
   a. 1 percent
   b. 10 percent
   c. 20 percent
   d. 30 percent

____ 10. Two of the most important resources in Washington and Oregon are
   a. wheat and dairy products.
   b. fish and forests.
   c. petroleum and natural gas.
   d. steel and iron.

Canada

True/False • *(10 points each)* Indicate whether each statement below is true or false by writing *T* or *F* in the space provided. If the statement is false, use the space below to explain why.

_____ **1.** Canada does not share a border with the United States.

_____ **2.** Canada shares some major physical regions with the United States.

_____ **3.** The Great Lakes are Lake Erie, Lake Michigan, Lake Huron, Lake Superior, and Lake Ontario.

_____ **4.** The St. Lawrence River links the Great Lakes to the Pacific Ocean.

_____ **5.** The Great Bear is one of Canada's largest lakes.

_____ **6.** The mildest part of Canada is in the far northwest.

_____ **7.** Canada contains no permafrost.

_____ **8.** The trees in Canada provide lumber and pulp.

_____ **9.** The United States, the United Kingdom, and Japan get much of their newsprint from Canada.

_____ **10.** Minerals are the least valuable of Canada's natural resources.

Canada

Fill in the Blank • *(10 points each)* For each of the following statements, fill in the blank with the appropriate word, phrase, or name.

1. The first Europeans to settle in Canada probably were the _____, who arrived around A.D. 1000.

2. _____ was the first European country to successfully settle parts of what would become Canada.

3. New France lasted for a century and a half before it was conquered by the

_____.

4. Almost a quarter of present-day Canadians are of _____ ancestry.

5. The Seven Years' War, which was fought from 1756 to 1763, was called the

_____ War in North America.

6. _____ are the administrative divisions of Canada.

7. Like the United States, Canada has a(n) _____ form of government.

8. Most of the people of what became known as the Northwest Territories were

Canadian Indians and _____, people of mixed European and native ancestry.

9. Many _____ people came to Canada to help build the railroads.

10. Canada's largest city today is _____.

Canada

SECTION 3

Matching • *(10 points each)* Match each of the following terms or places with the correct description by writing the letter of the description in the space provided. Some descriptions will not be used.

_____ **1.** regionalism

_____ **2.** Nova Scotia

_____ **3.** Quebec

_____ **4.** Ontario

_____ **5.** Alberta

_____ **6.** Edmonton

_____ **7.** British Columbia

_____ **8.** Vancouver

_____ **9.** Nunavut

_____ **10.** Inuit

a. Prairie Province in which farming and oil and natural gas production are important

b. Canada's westernmost province

c. Strong connection that people feel toward their region

d. Term that means "on or near the sea"

e. New territory that was created for the native Inuit who live there

f. Canada's most populous province

g. City that lies between Toronto and Montreal

h. One of the Maritime Provinces

i. Canada's main trade center with the Pacific Rim countries

j. Native people who live in the Canadian north

k. City whose older section has narrow streets, stone walls, and French-style architecture

l. One of the Prairie Provinces' major cities

Mexico

CHAPTER 8

Multiple Choice • *(10 points each)* For each of the following, write the letter of the *best* choice in the space provided.

_____ **1.** Mexico's Río Bravo is known in the United States as the
   a. Colorado River.
   b. Grand Canyon.
   c. Pacific Ocean.
   d. Rio Grande.

_____ **2.** A serious issue that the people in Mexico now face is
   a. lack of automobiles.
   b. volcanic eruptions.
   c. colliding tectonic plates.
   d. water scarcity.

_____ **3.** What underlies much of the Yucatán Peninsula?
   a. lava
   b. limestone
   c. ice
   d. quartz

_____ **4.** What causes the Mexican Plateau to experience surprisingly cool temperatures?
   a. its northern location
   b. the tropical rain forests
   c. volcanic activity
   d. the elevation

_____ **5.** The most valuable part of Mexico's mining industry is
   a. gold.   **c.** diamonds.
   b. silver.   **d.** platinum.

_____ **6.** Most of northern Mexico is
   a. arid.
   b. rainy.
   c. covered in ice.
   d. humid.

_____ **7.** What is the capital of Mexico?
   a. Tehuantepec
   b. Sierra Madre Occidental
   c. Mexico City
   d. Plateau of Mexico

_____ **8.** What is Mexico's most important mineral resource?
   a. diamonds
   b. natural gas
   c. petroleum
   d. coal

_____ **9.** Much of Mexico is made up of
   a. a central plateau.
   b. arid desert.
   c. a peninsula.
   d. deep sinkholes.

_____ **10.** Baja California separates the Gulf of California from the
   a. Atlantic Ocean.
   b. Gulf of Mexico.
   c. Pacific Ocean.
   d. Rio Grande.

Mexico

CHAPTER 8

True/False • *(10 points each)* Indicate whether each statement below is true or false by writing *T* or *F* in the space provided. If the statement is false, use the space below to explain why.

_____ **1.** The first people to live in Mesoamerica arrived from the north about 12,000 years ago.

_____ **2.** The Olmec built temples, pyramids, and huge statues.

_____ **3.** Modern scholars have been unable to read any Maya writing.

_____ **4.** The Aztec established their capital, Tenochtitlán, on an island in a lake in the Valley of Mexico.

_____ **5.** The native people of the Americas had no resistance to diseases brought by Europeans.

_____ **6.** Hernán Cortés and his conquistadores defeated the Maya.

_____ **7.** In New Spain, people of mixed European and Indian ancestry were known as mulattoes.

_____ **8.** Before the arrival of the Spaniards, Indian communities had owned and worked land in groups.

_____ **9.** In 1810 Miguel Hidalgo y Costilla, a Catholic priest, began a revolt against Spanish rule.

_____ **10.** In Mexico today one major indicator of ethnic group is religion.

Name _____ Class _____ Date _____

Mexico

Fill in the Blank • *(10 points each)* **For each of the following statements, fill in the blank with the appropriate word, phrase, or name.**

1. Mexico's government includes an elected president and a _____.

2. Since the 1980s Mexico has wrestled with debts to foreign banks, high unemployment, and _____.

3. _____ remains an important part of the Mexican economy, along with industry and tourism.

4. Because of high demand in the United States, Mexico has shifted to growing _____, or crops produced primarily to sell.

5. Only _____ percent of the land in Mexico can grow crops.

6. Greater _____ is Mexico's most developed and crowded region.

7. Mexico City is plagued by _____, which is a mixture of smoke, chemicals, and fog.

8. The population in the forested coastal plains between Tampico and Campeche has grown as _____ production there has increased.

9. Many foreign-owned factories, known as _____, are located in northern Mexico.

10. In the Yucatán Peninsula, some farmers clear areas of forest using a practice that is known as _____ agriculture.

Central America and the Caribbean Islands

CHAPTER 9

True/False • *(10 points each)* Indicate whether each statement below is true or false by writing *T* or *F* in the space provided. If the statement is false, use the space below to explain why.

_____ **1.** Central America forms a bridge between North America and South America.

_____ **2.** The Caribbean islands form an archipelago, or large group of peninsulas.

_____ **3.** The four large islands of the Greater Antilles are Cuba, Jamaica, Puerto Rico, and Hispaniola.

_____ **4.** The Bahamas are located west of Florida.

_____ **5.** The tectonic activity in the region of Central America is caused by colliding tectonic plates.

_____ **6.** Dense rain forests can be found along the Caribbean coast of Central America.

_____ **7.** A cloud forest is a high-elevation, very wet tropical forest where low clouds are common.

_____ **8.** Hurricanes are rare in the region of the Caribbean.

_____ **9.** Volcanic ash ruins soil and makes agriculture impossible in the region.

_____ **10.** Jamaica has large reserves of bauxite, the most important aluminum ore.

Central America and the Caribbean Islands

CHAPTER 9

SECTION 2

Fill in the Blank • *(10 points each)* For each of the following statements, fill in the blank with the appropriate word, phrase, or name.

1. In the early 1500s most of Central America came under the control of the country

   of _____.

2. The largest ethnic group in Central America is _____.

3. Central American countries have at times been ruled by _____, or people who rule a country with complete authority.

4. _____ is the most populous country in Central America.

5. The country with the smallest population is _____.

6. Only _____ percent of the land in Honduras is suitable for farming.

7. A(n) _____, which occurred in El Salvador, is a conflict between two or more groups within a country.

8. The largest country in Central America is _____, which has coasts on both the Caribbean Sea and the Pacific Ocean.

9. Costa Rica is well known for its _____, which is the practice of using an area's natural environment to attract tourists.

10. The _____ Canal links the Pacific Ocean to the Caribbean Sea and Atlantic Ocean.

Central America and the Caribbean Islands

CHAPTER 9

Matching • *(10 points each)* Match each of the following terms or places with the correct description by writing the letter of the description in the space provided. Some descriptions will not be used.

_____ **1.** Santería

_____ **2.** calypso

_____ **3.** refugees

_____ **4.** cooperatives

_____ **5.** Haiti

_____ **6.** Santo Domingo

_____ **7.** Puerto Rico

_____ **8.** commonwealth

_____ **9.** St. Kitts and Nevis

_____ **10.** plantains

a. National music and dance of the Dominican Republic

b. Poorest country in the Americas

c. Self-governing territory associated with another country

d. People who flee their homeland to go to another country

e. Haiti's capital and center of industry

f. Type of bananas used in cooking

g. Religion that began in Cuba and spread to nearby islands and parts of the United States

h. Commonwealth of the United States that was once a Spanish colony

i. Capital of the Dominican Republic

j. Smallest country in the Lesser Antilles

k. Organizations owned by their members and operated for their mutual benefit

l. Music that has Trinidad and Tobago as its home

Caribbean South America

SECTION 1

Fill in the Blank • *(10 points each)* For each of the following statements, fill in the blank with the appropriate word, phrase, or name.

1. In the western part of Caribbean South America, the _____ rise above 18,000 feet.

2. A(n) _____ is a mountain system made up of parallel ranges.

3. The _____ Highlands in the eastern part of the region have been eroding for millions of years.

4. The term _____ refers to sandstone layers that cap steep-sided plateaus.

5. The plains of the Orinoco River basin are the _____ of eastern Colombia and western Venezuela.

6. The _____ is the longest river in Caribbean South America.

7. Aggressive, meat-eating fish called _____ live in the Orinoco River.

8. The climates of the Andes are divided by _____ into five zones.

9. One typical crop that is grown in the *tierra templada,* or "temperate country," is

   _____ .

10. Good _____ and moderate climates help make Caribbean South America a rich agricultural area.

Caribbean South America

True/False • *(10 points each)* Indicate whether each statement below is true or false by writing *T* or *F* in the space provided. If the statement is false, use the space below to explain why.

_____ **1.** Some giant mounds of earth, stone statues, and tombs that have been found in Colombia are more than 1,500 years old.

_____ **2.** A Chibcha custom inspired the legend of El Dorado, or "the Golden One."

_____ **3.** When Spanish explorers arrived on the Caribbean coast around 1500, they were conquered by the Chibcha.

_____ **4.** The Republic of Gran Colombia included all of South America and North America.

_____ **5.** In 1830 New Granada became Venezuela.

_____ **6.** Today, Colombia is Caribbean South America's most populous country.

_____ **7.** The national capital of Colombia is Magdelena.

_____ **8.** Colombia is world-famous for its tea, which is grown in the country's rich soil.

_____ **9.** In recent years, Colombia's leading export has been oil.

_____ **10.** Colombia's main religion is Roman Catholicism.

Caribbean South America

Matching • *(10 points each)* Match each of the following terms, names, or places with the correct description by writing the letter of the description in the space provided. Some descriptions will not be used.

_____ **1.** Venezuelan coast

_____ **2.** indigo

_____ **3.** Simón Bolívar

_____ **4.** caudillo

_____ **5.** oil

_____ **6.** Caracas

_____ **7.** Maracaibo

_____ **8.** Guiana Highlands

_____ **9.** *llaneros*

_____ **10.** *pardos*

a. Venezuelan lake that is a bay of the Caribbean Sea and a rich source of oil

b. Person who led the revolt against Spain and who is now considered a hero in many South American countries

c. People of mixed African, European, and South American Indian ancestry who live in Venezuela

d. Resource that caused the Venezuelan economy to suffer when prices for the resource dropped

e. Columbus landed there in 1498

f. Percentage of Venezuelans who live in cities and towns

g. Person who landed on the Venezuelan coast in 1498

h. Capital of Venezuela

i. Cowboys of the Venezuelan plains

j. Plant used to make a deep blue dye

k. Area in southeast Venezuela that is rich in iron ore for making steel

l. Term for a Venezuelan military leader

CHAPTER 10 Caribbean South America

Multiple Choice • *(10 points each)* For each of the following, write the letter of the *best* choice in the space provided.

____ **1.** The first Europeans to claim the Guianas were the
   a. Dutch.
   b. Spanish.
   c. Portuguese.
   d. French.

____ **2.** During the early history of the Guianas, what became the main crop?
   a. sweet potatoes
   b. molasses
   c. corn
   d. sugarcane

____ **3.** Colonists in the Guianas brought in indentured servants from
   a. India, China, and Southeast Asia.
   b. Brazil and Venezuela.
   c. Greece and Turkey.
   d. Britain, France, and Spain.

____ **4.** British Guiana gained independence in 1966 and became known as
   a. Great Britain.
   b. Guyana.
   c. French Guiana.
   d. Suriname.

____ **5.** Guyana's major mineral resource is
   a. silver.
   b. iron.
   c. bauxite.
   d. gold.

____ **6.** The capital of Guyana is
   a. Caracas.
   b. Netherlands.
   c. British Guiana.
   d. Georgetown.

____ **7.** Many farms in Suriname are found
   a. in coastal areas.
   b. in the interior plains.
   c. in the mountains.
   d. near the desert.

____ **8.** A major export in Suriname is
   a. aluminum.
   b. bauxite.
   c. diamonds.
   d. oil.

____ **9.** French Guiana's most important economic activities are
   a. oil production and steelmaking.
   b. forestry and shrimp fishing.
   c. agriculture and shipbuilding.
   d. cheese making and wine making.

____ **10.** About two thirds of the people who live in French Guiana are descended from
   a. Vikings.
   b. Normans.
   c. Africans.
   d. Celts.

Atlantic South America

Fill in the Blank • *(10 points each)* For each of the following statements, fill in the blank with the appropriate word, phrase, or name.

1. Atlantic South America includes four countries—Brazil, _____, Uruguay, and Paraguay.

2. The _____ River basin that is located in northern Brazil is a giant, flat floodplain.

3. The Gran _____ is a lower region that is south of the Brazilian Plateau.

4. The highest mountains in South America are the _____.

5. The largest river system in the world, the _____, flows across northern Brazil.

6. The Río de la Plata is a(n) _____, a partially enclosed body of water where salty seawater and freshwater mix.

7. The humid tropical climate of the Amazon River basin supports the world's

   largest tropical _____, where it rains almost every day.

8. Because it has rich soils and a humid subtropical climate, the _____ is a major farming region.

9. In some areas of Atlantic South America, planting the same crop every year has caused

   soil _____, meaning that the soil has lost nutrients needed by plants.

10. One of the largest _____ dams in the world is the Itaipu Dam on the Paraná River between Brazil and Paraguay.

Atlantic South America

Multiple Choice • *(10 points each)* For each of the following, write the letter of the *best* choice in the space provided.

_____ 1. Brazilian Indians are descended from people who probably came from
   a. Asia.
   b. Argentina.
   c. Nicaragua.
   d. the Netherlands.

_____ 2. Brazilians use cassava
   a. to mine for gold and silver.
   b. as their main form of transportation.
   c. to build their homes.
   d. as an ingredient in foods.

_____ 3. Which of the following groups began to move into Brazil after the 1500s?
   a. the French   c. the Spanish
   b. the British   d. the Portuguese

_____ 4. Colonists eventually replaced forests along the Atlantic coast with
   a. pineapple plantations.
   b. sugar plantations.
   c. secondary forests.
   d. sheep ranches.

_____ 5. What is the major religion in Brazil?
   a. *Macumba*
   b. Protestantism
   c. Roman Catholicism
   d. Islam

_____ 6. Many Brazilians eat *feijoada,* a stew of black beans and
   a. meat.
   b. rice.
   c. red peppers.
   d. seafood.

_____ 7. Which of the following is Brazil's poorest region?
   a. the Amazon
   b. the northeast
   c. the southeast
   d. the interior

_____ 8. What are favelas?
   a. small hills
   b. insects of the rain forest
   c. huge slums
   d. beans grown only in Brazil

_____ 9. The richest region in Brazil is
   a. the Amazon.
   b. the northeast.
   c. the southeast.
   d. the interior.

_____ 10. The present-day capital of Brazil is
   a. Rio de Janeiro.
   b. Belém.
   c. São Paulo.
   d. Brasília.

Atlantic South America

CHAPTER 11

SECTION 3

Matching • *(10 points each)* Match each of the following terms or places with the correct description by writing the letter of the description in the space provided. Some descriptions will not be used.

_____ **1.** Argentina

_____ **2.** *encomienda*

_____ **3.** Pampas

_____ **4.** gaucho

_____ **5.** United Kingdom

_____ **6.** Spanish

_____ **7.** *parrillada*

_____ **8.** agriculture

_____ **9.** Buenos Aires

_____ **10.** Mercosur

a. Combines the Spanish tango with a dance called *milonga*

b. Argentine cowboy

c. Capital of Argentina

d. Country with which Argentina fought a brief war in 1983

e. Main religion in Argentina

f. Name that means "land of silver" or "silvery one"

g. Official language of Argentina

h. Part of the Argentine economy in which about 12 percent of the labor force works

i. System under which the Spanish monarch gave colonists land and the right to the labor of Indians who lived on that land

j. Trade organization that promotes economic cooperation among its members in southern and eastern South America

k. Important agricultural region during Argentina's colonial era

l. Sausages and steaks served on a small grill

Atlantic South America

True/False • *(10 points each)* Indicate whether each statement below is true or false by writing *T* or *F* in the space provided. If the statement is false, use the space below to explain why.

_____ **1.** The capital of Uruguay is Paraguay.

_____ **2.** People of European descent make up about 88 percent of Uruguay's population.

_____ **3.** The main religion of Uruguay is Protestantism.

_____ **4.** Only about 10 percent of Uruguayans live in urban areas.

_____ **5.** An important source of energy in Uruguay is hydroelectric power.

_____ **6.** Paraguay shares borders with Bolivia, Brazil, and Argentina, and is a landlocked country.

_____ **7.** About 95 percent of Paraguayans are Indians.

_____ **8.** Spanish is the official language of Paraguay.

_____ **9.** The capital of Paraguay and its largest city is Asunción.

_____ **10.** Nearly half of the workers in Paraguay are small-business owners.

Pacific South America

SECTION 1

Matching • *(10 points each)* Match each of the following terms or places with the correct description by writing the letter of the description in the space provided. Some descriptions will not be used.

_____ **1.** Ecuador

_____ **2.** Bolivia

_____ **3.** Andes

_____ **4.** strait

_____ **5.** Tierra del Fuego

_____ **6.** *selvas*

_____ **7.** Atacama Desert

_____ **8.** Peru Current

_____ **9.** Chile

_____ **10.** El Niño

a. Narrow passageway that connects two large bodies of water

b. Pacific South American country whose coastal area is one of the cloudiest and driest places on Earth

c. Pacific South American country that is landlocked

d. Broad, high-plateau plain located between southern Peru and Bolivia

e. Area where rain is extremely rare but where fog and low clouds are common

f. Pacific South American country that forms a long curve along the Pacific Ocean

g. South American name for the thick tropical rain forests in the region

h. Chills the warmer air above the ocean's surface and causes fog and low clouds to form

i. Ocean and weather pattern that affects the dry Pacific coast every two to seven years

j. Pacific South American country that lies on the equator

k. Large island south of the Strait of Magellan

l. Snowcapped mountains that run through all four countries in the Pacific South American region

Pacific South America

True/False • *(10 points each)* Indicate whether each statement below is true or false by writing *T* or *F* in the space provided. If the statement is false, use the space below to explain why.

_____ **1.** The early people in Pacific South America domesticated quinoa, which is a breed of dog.

_____ **2.** The Tiahuanaco culture left behind huge stone carvings near the Bolivian shores of Lake Titicaca.

_____ **3.** The Inca civilization stretched from the Pacific Coast inland to the *selvas* of the Amazon rain forest and contained as many as 12 million people.

_____ **4.** The Inca's greatest achievement was the development of radio.

_____ **5.** The Inca were defeated by the Spanish explorer Francisco Pizarro and his soldiers.

_____ **6.** The first country in Pacific South America to achieve independence from Spain was Bolivia.

_____ **7.** Since they gained their independence, all the countries of Pacific South America have had periods during which their governments were unstable.

_____ **8.** The Pacific South American country of Ecuador is now a democracy.

_____ **9.** The country of Peru has an elected president and congress.

_____ **10.** Since it declared independence from Spain, Chile has experienced only peace.

Name _____ Class _____ Date _____

Pacific South America

SECTION 3

Fill in the Blank • *(10 points each)* For each of the following statements, fill in the blank with the appropriate word, phrase, or name.

1. Quito, the capital of _____, is located in the Andean region.

2. Although Spanish is the official language of Ecuador, about 19 percent of the people speak native languages, such as _____, the language of the Inca.

3. The country of Bolivia has two capitals: _____ and Sucre.

4. Bolivia has the highest percentage of _____ of any South American country.

5. _____, like Ecuador, has three major regions.

6. Stone structures dating from the Inca period draw many tourists to places like Cuzco and _____.

7. The capital of Peru is _____, which is served by the seaport city of Callao.

8. A(n) _____ is a group of military officers who rule a country after seizing power.

9. _____ mining accounts for over a third of Chile's exports.

10. Some people have suggested that Chile should be the next country to join _____, the free-trade group that includes Canada, Mexico, and the United States.

Southern Europe

Multiple Choice • *(10 points each)* For each of the following, write the letter of the *best* choice in the space provided.

_____ **1.** Southern Europe is also known as
  a. Mesoamerica.
  b. Mediterranean Europe.
  c. the Strait of Gibraltar.
  d. the Cantabrian.

_____ **2.** What links the Mediterranean to the Atlantic Ocean?
  a. the Pacific Ocean
  b. the Gulf of Mexico
  c. the Strait of Magellan
  d. the Strait of Gibraltar

_____ **3.** How many peninsulas make up southern Europe?
  a. two
  b. three
  c. five
  d. seven

_____ **4.** Which of the following can be found on the Iberian Peninsula?
  a. Spain and Portugal
  b. Greece
  c. France and Britain
  d. Sweden, Norway, and Finland

_____ **5.** Which of the following is the largest river in Italy?
  a. Ebro
  b. Po
  c. Tiber
  d. Douro

_____ **6.** The largest island in Greece is
  a. Crete.
  b. Sardinia.
  c. Gibraltar.
  d. Peloponnesus.

_____ **7.** Most rain in southern Europe falls during
  a. fall.
  b. spring.
  c. winter.
  d. summer.

_____ **8.** Which of the following describes the climate of much of southern Europe?
  a. cold and snowy
  b. cool and rainy
  c. hot and arid
  d. warm and sunny

_____ **9.** The siroccos that blow over Italy originate from
  a. North America.
  b. Asia.
  c. North Africa.
  d. South America.

_____ **10.** Which of the following countries mines bauxite, chromium, lead, and zinc?
  a. Portugal
  b. Spain
  c. Greece
  d. Italy

Southern Europe

SECTION 2

True/False • *(10 points each)* Indicate whether each statement below is true or false by writing *T* or *F* in the space provided. If the statement is false, use the space below to explain why.

_____ **1.** By about 2000 B.C. Crete had large towns and a complex civilization.

_____ **2.** Each Greek city-state was made up of a city and the land around it.

_____ **3.** The government of Macedonia was the first known democracy.

_____ **4.** Alexander the Great conquered Asia Minor, Egypt, Persia, and part of India.

_____ **5.** The Byzantine Empire was ruled from Rome.

_____ **6.** The Ottoman Turks conquered Constantinople in 1453.

_____ **7.** During World War II, Greece was occupied by Japan.

_____ **8.** Ninety-eight percent of Greeks are Eastern Orthodox Christians, commonly known as Greek Orthodox.

_____ **9.** The ancient Greeks made mosaics that were copied throughout Europe.

_____ **10.** In Greece today, more people work in agriculture than in any other industry.

Southern Europe

Fill in the Blank • *(10 points each)* **For each of the following statements, fill in the blank with the appropriate word, phrase, or name.**

1. The Romans made advances in engineering, including roads and _____, canals that transport water.

2. The western part of the Roman Empire, with its capital in _____, fell in A.D. 476.

3. _____ developed into the modern languages of French, Italian, Portuguese, Romanian, and Spanish.

4. _____ became the main religion of the Roman Empire when it was adopted by Emperor Constantine in the early A.D. 300s.

5. The head of the Roman Catholic Church is the _____, the bishop of Rome.

6. Beginning in the 1300s a new era of learning, called the _____, began in Italy.

7. During this new era of learning, Italians rediscovered the ancient cultures of Greece

   and _____.

8. An Italian named _____ perfected the telescope and experimented with gravity.

9. Ninety-eight percent of Italians belong to the _____ Church.

10. Italy's most valuable crop today is _____.

Southern Europe

Matching • (10 points each) Match each of the following terms, names, or places with the correct description by writing the letter of the description in the space provided. Some descriptions will not be used.

_____ **1.** Portugal

_____ **2.** Pablo Picasso

_____ **3.** Moors

_____ **4.** Christopher Columbus

_____ **5.** Philip II

_____ **6.** Castilian

_____ **7.** Basque

_____ **8.** fiesta

_____ **9.** Lisbon

_____ **10.** Madrid

a. King of Spain and Portugal who sent the Spanish Armada to invade England

b. One of two countries situated on the Iberian Peninsula

c. Portugal's capital and largest city

d. Festival that honors a patron saint

e. Old Moorish capital, which is a city in southern Spain

f. Language spoken in Catalonia in northeastern Spain

g. Ethnic group that uses violence to protest Spanish control

h. Spanish painter who became world famous

i. Muslim North Africans who conquered most of the Iberian Peninsula in the A.D. 700s

j. Most widely understood Spanish dialect

k. Spain's capital and largest city

l. Person sponsored by King Ferdinand and Queen Isabella to sail to the Americas

West-Central Europe

CHAPTER 14

SECTION 1

True/False • *(10 points each)* Indicate whether each statement below is true or false by writing *T* or *F* in the space provided. If the statement is false, use the space below to explain why.

_____ **1.** The Benelux countries are Switzerland and Austria.

_____ **2.** Belgium, the Netherlands, and Luxembourg are sometimes called the Low Countries.

_____ **3.** The landforms of west-central Europe are arranged like a fan.

_____ **4.** The Schwarzwald is also known as the White Forest.

_____ **5.** The highest mountain range in Europe is the Alps.

_____ **6.** The Alps have a marine west coast climate.

_____ **7.** The Seine, the Loire, the Garonne, and the Rhone Rivers can be found in France.

_____ **8.** West-central Europe has hundreds of excellent harbors.

_____ **9.** Germany and France both produce grapes for some of the world's finest wines.

_____ **10.** Alpine rivers provide geothermal power to Switzerland and Austria.

West-Central Europe

SECTION 2

Multiple Choice • *(10 points each)* For each of the following, write the letter of the *best* choice in the space provided.

____ **1.** After the Roman Empire collapsed, much of Gaul was taken over by the
   a. Vikings.
   b. Franks.
   c. British.
   d. Italians.

____ **2.** Who ruled over much of western Europe after the fall of the Roman Empire?
   a. Napoléon Bonaparte
   b. Cézanne
   c. Charlemagne
   d. Debussy

____ **3.** The area now known as Normandy was originally settled by Normans from
   a. Scandinavia.
   b. Australia.
   c. Africa.
   d. northern Europe.

____ **4.** The period from the fall of the Roman Empire to about 1500 is known as
   a. the Middle Ages.
   b. the Dark Time.
   c. the Era of Great Changes.
   d. Time's Arrow.

____ **5.** Which of the following is the largest city in France?
   a. Nice
   b. Marseille
   c. Paris
   d. Lille

____ **6.** The French Revolution began in the year
   a. 1776.
   b. 1789.
   c. 1812.
   d. 1860.

____ **7.** What is NATO?
   a. the new European currency
   b. an organization of European grape growers
   c. a military alliance created to defend Western Europe
   d. the head of the French government

____ **8.** About 90 percent of French people are
   a. Roman Catholic.
   b. Basque.
   c. Jewish.
   d. Protestant.

____ **9.** In the late 1800s and early 1900s France was the center of an artistic movement known as
   a. expressionism.
   b. hip-hop.
   c. realism.
   d. impressionism.

____ **10.** The Hundred Years' War began when which king tried to claim the throne of France?
   a. the king of Italy
   b. the king of Germany
   c. the king of Russia
   d. the king of England

West-Central Europe

SECTION 3

Fill in the Blank • *(10 points each)* For each of the following statements, fill in the blank with the appropriate word, phrase, or name.

1. When the Roman Empire collapsed, the _____ became the most important tribe in Germany.

2. Charlemagne's empire was known as the _____ Empire.

3. During the 1500s Germany was the center of an effort to reform Christianity, an effort known as the _____.

4. During the late 1800s _____, the strongest German state, led the creation of a united Germany.

5. _____ established a new political party, the Nazis, and came to power in Germany in 1933.

6. The killing of millions of Jews and other people by the Nazis during World War II is called the _____.

7. In 1989 the _____ separating East and West Germany was torn down, and in 1990 the two countries were reunited.

8. The major German festival season is _____.

9. Germany's capital city, _____, has wide boulevards and many parks.

10. Near the Rhine River lies a huge cluster of cities that form Germany's largest industrial district, called the _____.

West-Central Europe

SECTION 4

True/False • *(10 points each)* Indicate whether each statement below is true or false by writing *T* or *F* in the space provided. If the statement is false, use the space below to explain why.

_____ **1.** Celtic and Germanic tribes once lived in this region.

_____ **2.** In the 1570s the Protestants of the Netherlands won their freedom from German rule.

_____ **3.** Most of the major battles of World War I were fought in Luxembourg.

_____ **4.** Each of the three Benelux countries is ruled by a parliament and a monarch.

_____ **5.** The people who live in Luxembourg and Belgium are mostly Protestant.

_____ **6.** The main language of the Netherlands is English.

_____ **7.** The English word "cookies" comes from the Dutch word "Koekjes."

_____ **8.** The Netherlands is famous for its flowers, especially its roses.

_____ **9.** Today the country of Luxembourg earns much of its income from the provision of services such as banking.

_____ **10.** The headquarters of the European Union and of NATO can be found in Brussels.

West-Central Europe

Matching • *(10 points each)* Match each of the following terms, names, or places with the correct description by writing the letter of the description in the space provided. Some descriptions will not be used.

_____ **1.** Mozart

_____ **2.** canton

_____ **3.** Switzerland

_____ **4.** Austria

_____ **5.** nationalism

_____ **6.** Napoléon Bonaparte

_____ **7.** Roman Catholic

_____ **8.** German

_____ **9.** Christmas

_____ **10.** Vienna

a. The demand for self-rule

b. Dominant language of Austria

c. Wrote symphonies and operas in Salzburg

d. The center of banking in Switzerland

e. Austria's commercial and industrial center

f. Country that was the home of the Habsburgs, a powerful family of German nobles

g. Dominant religion of Austria

h. Term for a Swiss district

i. Person whose defeat allowed the Austrian Empire to become the dominant power in central Europe

j. The capital city of Switzerland

k. Country that gained its independence in the 1600s after its districts gradually broke away from the Holy Roman Empire

l. Major festival in both Austria and Switzerland

Northern Europe

True/False • *(10 points each)* Indicate whether each statement below is true or false by writing *T* or *F* in the space provided. If the statement is false, use the space below to explain why.

_____ **1.** Iceland is the world's largest island.

_____ **2.** Scandinavia is made up of Denmark, Sweden, Finland, Iceland, and Norway.

_____ **3.** Greenland is mostly covered by rolling green hills.

_____ **4.** Many fjords can be found in Norway.

_____ **5.** The longest river in the British Isles is the Shannon River, which can be found in Ireland.

_____ **6.** Special ships keep the sea lanes to Sweden and Finland open.

_____ **7.** The farms in northern Europe grow many cool-climate crops.

_____ **8.** Iceland uses geothermal and hydroelectric power to supply its energy needs.

_____ **9.** Much of the northern European region has a desert climate.

_____ **10.** Central Sweden and southern Finland have four true seasons.

Northern Europe

Multiple Choice • *(10 points each)* For each of the following, write the letter of the *best* choice in the space provided.

_____ **1.** The last people to conquer the British Isles were the
  a. Celts.
  b. Normans.
  c. Saxons.
  d. Vikings.

_____ **2.** What is the largest city in the United Kingdom today?
  a. London
  b. Belfast
  c. Glasgow
  d. Manchester

_____ **3.** Three early industries in the United Kingdom were
  a. automaking, shipbuilding, and textiles.
  b. shipbuilding, nuclear power, and fishing.
  c. automaking, iron and steel, and oil refining.
  d. textiles, shipbuilding, and iron and steel.

_____ **4.** Laws in the United Kingdom are made by
  a. a queen.
  b. a parliament.
  c. a prime minister.
  d. a president.

_____ **5.** Which of the following is the official language of the United Kingdom?
  a. Spanish
  b. French
  c. English
  d. Gaelic

_____ **6.** The popular meal of fish and chips consists of fried fish and
  a. potatoes.
  b. tomatoes.
  c. corn chips.
  d. peas.

_____ **7.** What percentage of the world's land area did the British Empire control in 1900?
  a. 25 percent
  b. 50 percent
  c. 75 percent
  d. 100 percent

_____ **8.** One area of Scotland is called Silicon Glen because of its
  a. important textile industry.
  b. numerous glass-making shops.
  c. ceramics factories.
  d. computer and electronics businesses.

_____ **9.** What percentage of the labor force in the United Kingdom works in agriculture?
  a. 1 percent
  b. 10 percent
  c. 25 percent
  d. 60 percent

_____ **10.** Which of the following has been the source of much violence in Northern Ireland?
  a. race
  b. land
  c. religion
  d. ethnicity

Northern Europe

Fill in the Blank • *(10 points each)* For each of the following statements, fill in the blank with the appropriate word, phrase, or name.

1. The country of _____ conquered Ireland in the A.D. 1100s.

2. English and Irish _____, a Celtic language, are the official languages of Ireland.

3. In the 1840s a poor economy and a potato _____ forced millions of Irish people to leave Ireland for other countries, including the United States.

4. All ties between the _____ of Ireland and the British Empire were cut in 1949.

5. Much of Ireland is covered with _____, soft ground soaked with water.

6. Music is so important in Ireland that the Irish _____ is a national symbol.

7. More than 90 percent of the people living in Ireland today are members of the

   _____ religion.

8. St. _____ is believed to have brought Christianity to Ireland in the 400s, and in his honor March 17 is a national holiday in Ireland.

9. Ireland's membership in the _____ has greatly helped its economy.

10. Ireland's capital and largest city is _____.

Northern Europe

CHAPTER 15

Matching • *(10 points each)* Match each of the following terms or places with the correct description by writing the letter of the description in the space provided. Some descriptions will not be used.

_____ **1.** Finnish

_____ **2.** Oslo

_____ **3.** Sweden

_____ **4.** neutral

_____ **5.** Denmark

_____ **6.** Greenland

_____ **7.** uninhabitable

_____ **8.** Iceland

_____ **9.** Finland

_____ **10.** Lapland

a. Territory of Denmark that is a part of North America

b. Hot springs that shoot hot water and steam into the air

c. Only Scandinavian language that is not closely related to all the other Scandinavian languages

d. Term used to describe the decision not to take sides in a conflict

e. Smallest and most densely populated of the Scandinavian countries

f. Cultural region populated by Sami, who herd reindeer

g. Capital of Norway and one of its largest cities

h. Easternmost country in the Scandinavian region

i. Capital of Sweden

j. Country whose capital and largest city is Reykjavik

k. Largest and most populous country in Scandinavia

l. Term used to describe an area that cannot support human settlement

Name _____ Class _____ Date _____

Eastern Europe

SECTION 1

True/False • *(10 points each)* Indicate whether each statement below is true or false by writing *T* or *F* in the space provided. If the statement is false, use the space below to explain why.

_____ **1.** The Baltic countries are Estonia, Latvia, and Lithuania.

_____ **2.** Eastern Europe is a region of mountains and plains.

_____ **3.** The Carpathian Mountains stretch from the Czech Republic across southern Poland and Slovakia and into Ukraine.

_____ **4.** Eastern Europe's most important river for trade and transportation is the Transylvanian River.

_____ **5.** The eastern half of the region has long, snowy winters and short, rainy summers.

_____ **6.** Layered rock that produces oil when heated is called quartz.

_____ **7.** Slovenia and Slovakia mine lignite, a soft form of coal.

_____ **8.** The people who live in central Poland today mine salt, a practice that began in Poland in the 1200s.

_____ **9.** Fossilized tree sap is called amber.

_____ **10.** The region of Eastern Europe does not have the pollution problems that trouble other areas of the world.

Eastern Europe

SECTION 2

Matching • *(10 points each)* Match each of the following terms, places, or people with the correct description by writing the letter of the description in the space provided. Some descriptions will not be used.

_____ **1.** Velvet Revolution

_____ **2.** Balts

_____ **3.** Huns

_____ **4.** Mongols

_____ **5.** Czechoslovakia

_____ **6.** Prague

_____ **7.** Estonia

_____ **8.** Poland

_____ **9.** Slovakia

_____ **10.** Hungary

a. People who rode out of Central Asia into Hungary in the 1200s

b. Present-day capital of the Czech Republic

c. Country whose manufacturing industry is located in and around Budapest, the country's capital

d. Term that refers to the collapse of communist rule in Czechoslovakia in 1989

e. Largest of the Baltic countries

f. Country that along with the Czech Republic was created out of Czechoslovakia in 1993

g. Country whose capital is Vilnius

h. Northeastern Europe's largest and most populous country

i. Ancient, Indo-European-speaking people who gave their name to the Baltic Sea

j. Country whose language is related to Finnish

k. Group of nomadic warriors who invaded Eastern Europe in the A.D. 400s

l. Country that along with Yugoslavia was created by the peace treaty that ended World War I

Eastern Europe

Fill in the Blank • *(10 points each)* For each of the following statements, fill in the blank with the appropriate word, phrase, or name.

1. The _____ ruled southeastern Europe until the 1800s.

2. When the Austro-Hungarians took control of Croatia and Slovenia in the 1800s,

   they imposed the _____ religion there.

3. In 1914, when a Serb nationalist killed the heir to the Austro-Hungarian throne,

   Austria declared war on Serbia and _____ began.

4. The _____ are the most diverse region of Europe in terms of language, ethnicity, and religion.

5. Belgrade, located on the _____ River, is the capital of both Serbia and Yugoslavia.

6. Today, more people in Romania work in _____ than in any other part of the economy.

7. _____, or Gypsies as they were once known, are descended from people who may have lived in northern India centuries ago.

8. Macedonia's government has tried to move the country toward a _____ economy, or an economy in which consumers help determine what is to be produced.

9. The country of _____ has mainly an agricultural economy, with most industries located near Sofia, the capital and largest city.

10. _____, with its capital at Tiranë, is one of Europe's poorest countries.

Russia

CHAPTER 17

True/False • *(10 points each)* Indicate whether each statement below is true or false by writing *T* or *F* in the space provided. If the statement is false, use the space below to explain why.

_____ **1.** Russia was the smallest republic of what used to be called the Union of Soviet Socialist Republics.

_____ **2.** The Caucasus Mountains stretch from the Black Sea to the Caspian Sea.

_____ **3.** Some of the world's longest rivers, including the Volga and Don Rivers, flow through Russia.

_____ **4.** The longest river in Europe is the Amur River.

_____ **5.** Nearly all of Russia is located at low southern latitudes.

_____ **6.** Winters in Siberia are very difficult, with temperatures sometimes dropping below minus 40°F.

_____ **7.** The vast grasslands of the steppe stretch from Ukraine across southern Russia to Kazakhstan.

_____ **8.** Russia has few energy, mineral, and forest resources.

_____ **9.** The Russian taiga provides a huge supply of trees for wood and paper pulp.

_____ **10.** Russia is a major producer of diamonds.

Russia

Multiple Choice • *(10 points each)* For each of the following, write the letter of the *best* choice in the space provided.

_____ 1. The Rus, who helped shape the first Russian state, actually were from
   a. France.
   b. North America.
   c. Spain.
   d. Scandinavia.

_____ 2. Today, Kiev is the capital of
   a. Ukraine.
   b. Moscow.
   c. Russia.
   d. Muscovy.

_____ 3. What alphabet does the Russian language use today?
   a. Latin
   b. English
   c. Cyrillic
   d. Greek

_____ 4. By the early 1700s the Russian Empire stretched from the Baltic Sea to the
   a. Pacific Ocean.
   b. Gulf of Mexico.
   c. Mediterranean Sea.
   d. Atlantic Ocean.

_____ 5. In 1867 Russia sold Alaska to
   a. Germany.
   b. the United States.
   c. Canada.
   d. Great Britain.

_____ 6. During World War II, the United States and the Soviet Union were allies in the fight against
   a. France.
   b. Germany.
   c. Scandinavia.
   d. Great Britain.

_____ 7. The Cold War was a rivalry between the Soviet Union and
   a. Great Britain.
   b. Germany.
   c. Siberia.
   d. the United States.

_____ 8. What is black caviar?
   a. goat cheese
   b. a type of cake
   c. fish eggs
   d. a preserve made of blackberries

_____ 9. Aleksandr Solzhenitsyn is a Russian
   a. writer.
   b. astronaut.
   c. president.
   d. painter.

_____ 10. The Russian Federation is governed by an elected president and a
   a. queen.
   b. prime minister.
   c. chancellor.
   d. legislature.

Russia

SECTION 3

Fill in the Blank • *(10 points each)* **For each of the following statements, fill in the blank with the appropriate word, phrase, or name.**

1. The _____ section of Russia is the heartland of the country.

2. Farmers in the plains of European Russia focus on growing _____ and raising livestock.

3. _____ is Russia's capital and largest city.

4. The red brick walls and towers of the _____, the symbol of Russian and Soviet power, were built in the late 1400s.

5. _____ industry focuses on the production of lightweight goods, such as clothing.

6. _____ served as Russia's capital and the home of the czars for more than 200 years, until 1918.

7. The _____ region stretches along the middle part of the Volga River.

8. During World War II, many Soviet factories were moved to the Volga region to keep them safe from _____ invaders.

9. Factories that process metal ores are called _____.

10. Many large cities in the _____ region started as commercial centers for mining districts.

Russia

True/False • *(10 points each)* Indicate whether each statement below is true or false by writing *T* or *F* in the space provided. If the statement is false, use the space below to explain why.

_____ **1.** Siberia is west of European Russia.

_____ **2.** Siberia is nearly 1.5 times the size of the United States.

_____ **3.** During the winter months habitation fog, a fog caused by fumes and smoke from cities, hangs over Siberia.

_____ **4.** Siberia is densely populated.

_____ **5.** The Trans-Siberian Railroad is the longest single rail line in the world.

_____ **6.** Only a small percentage of Russia's industry can be found in Siberia.

_____ **7.** Fishing and steel-making are Siberia's most important industries.

_____ **8.** Large deposits of coal are mined in the Kuznetsk Basin, or Kuzbas.

_____ **9.** The name of Siberia's largest city, Novosibirsk, means "Old Siberia."

_____ **10.** The world's deepest lake is Lake Baikal, known as the "Diamond of Siberia."

Russia

CHAPTER 17

SECTION 5

Matching • *(10 points each)* Match each of the following terms or places with the correct description by writing the letter of the description in the space provided. Some descriptions will not be used.

_____ **1.** sable

_____ **2.** geothermal energy

_____ **3.** Trans-Siberian Railroad

_____ **4.** Khabarovsk

_____ **5.** Vladivostok

_____ **6.** icebreaker

_____ **7.** Sakhalin

_____ **8.** Kuril Islands

_____ **9.** Japan

_____ **10.** Pacific

a. City founded in 1858, where more than 600,000 people now live

b. Site of Russia's first geothermal electric-power station

c. Energy resource available in the Russian Far East because of the region's tectonic activity

d. Islands that stretch from Hokkaido to the Kamchatka Peninsula

e. Ship that can break up the ice of frozen waterways

f. River where Khabarovsk is located

g. Used to make expensive clothing

h. Large island that lies off the eastern coast of Siberia

i. Location of the Kurils and Sakhalin

j. City whose name means "Lord of the East" in Russian

k. Country with which Russia has had a disagreement over the ownership of Sakhalin and the Kurils

l. Aided the growth of cities in the Russian Far East

Ukraine, Belarus, and the Caucasus

True/False • *(10 points each)* Indicate whether each statement below is true or false by writing *T* or *F* in the space provided. If the statement is false, use the space below to explain why.

_____ **1.** The countries of Georgia, Armenia, and Azerbaijan lie in a rugged region called the Himalayas.

_____ **2.** The Caucasus region is located between the Black Sea and the Caspian Sea.

_____ **3.** The region's highest peak is Mount Elbrus.

_____ **4.** Earthquakes often occur south of the Caucasus.

_____ **5.** One of Europe's rivers, the Danube, flows south through Belarus and Ukraine.

_____ **6.** Ukraine has created several nature reserves to try to preserve its natural environments.

_____ **7.** The northern two thirds of Ukraine and Belarus has a marine west coast climate.

_____ **8.** A steppe climate can be found in Azerbaijan.

_____ **9.** Ukraine's greatest natural resource is rich farmland.

_____ **10.** Azerbaijan's large and valuable oil and gas deposits are found under the Mediterranean Sea.

Ukraine, Belarus, and the Caucasus

CHAPTER 18

SECTION 2

Matching • *(10 points each)* Match each of the following terms or places with the correct description by writing the letter of the description in the space provided. Some descriptions will not be used.

_____ **1.** Greeks

_____ **2.** Vikings

_____ **3.** Kiev

_____ **4.** Cyrillic

_____ **5.** Cossacks

_____ **6.** soviet

_____ **7.** Belarus

_____ **8.** Ukraine

_____ **9.** Chernobyl

_____ **10.** Minsk

a. Capital of present-day Ukraine

b. Administrative center of the Commonwealth of Independent States and the capital of Belarus

c. Nomadic horsemen who lived on the Ukrainian frontier

d. Town that in 1986 experienced one of the world's worst nuclear-reactor disasters

e. Person bound to the land who works for a lord

f. People who by about 600 B.C. had established trading colonies along the coast of the Black Sea

g. Council of a Soviet republic

h. Ukraine's most important food crop

i. Alphabet introduced to the Ukrainians and Belorussians by Greek Orthodox missionaries

j. Soviet republic that became a major industrial center

k. People who by the 800s had taken the city of Kiev

l. Country that has a good climate for growing crops and some of the world's richest soils

Ukraine, Belarus, and the Caucasus

SECTION 3

Fill in the Blank • *(10 points each)* For each of the following statements, fill in the blank with the appropriate word, phrase, or name.

1. In the 500s B.C. the Caucasus region was controlled by the ancient

_____ Empire.

2. In 1922 Georgia, Armenia, and Azerbaijan became part of the _____.

3. The countries of Georgia, Armenia, and Azerbaijan each have an elected

parliament, _____, and prime minister.

4. _____ is a small country located between the Caucasus Mountains and the Black Sea.

5. The country of Georgia must import most of its energy supplies because its only

energy resource is _____.

6. Armenia, which is a little smaller than the state of Maryland, lies just east of the

country of _____.

7. _____ accounts for about 40 percent of Armenia's gross domestic product.

8. The population of Azerbaijan is becoming more ethnically _____, or ethnically the same.

9. Azerbaijan is mainly a(n) _____ society, or a society organized around farming.

10. Azerbaijan's sturgeon roe, or _____, is made into some of the world's most sought-after caviar.

Central Asia

True/False • *(10 points each)* Indicate whether each statement below is true or false by writing *T* or *F* in the space provided. If the statement is false, use the space below to explain why.

_____ **1.** Central Asia lies in the middle of the largest continent.

_____ **2.** Central Asia lies north of some of the world's lowest mountain ranges.

_____ **3.** North of the Aral Sea, the rainfall is heavy enough for steppe vegetation.

_____ **4.** The Kara-Kum and the Kyzyl Kum are volcanoes in Central Asia.

_____ **5.** An oasis, several of which can be found in Central Asia, is a place in the desert where a spring or well provides water.

_____ **6.** The main water sources in southern Central Asia are the Syr Dar'ya and Amu Dar'ya Rivers.

_____ **7.** When it first flows down from the mountains, the Amu Dar'ya passes through the Fergana Valley.

_____ **8.** During the Soviet period, Central Asia's population grew slowly.

_____ **9.** In recent years, the Aral Sea has lost much of its water.

_____ **10.** The Central Asian economies' best chance for improvement lies in their fossil fuels.

Central Asia

Multiple Choice • *(10 points each)* For each of the following, write the letter of the *best* choice in the space provided.

____ **1.** Central Asia was the best land route for trade between China and
a. the Mediterranean Sea.
b. Japan.
c. Mexico.
d. South America.

____ **2.** Which of the following groups conquered Central Asia in the 1200s?
a. Spaniards
b. British
c. Mongols
d. Japanese

____ **3.** The Soviet Union set up five republics in Central Asia following
a. World War I.
b. the Russian Revolution.
c. the American Revolution.
d. World War II.

____ **4.** Soviet rule brought all of the following benefits to Central Asia except
a. hospitals.
b. freedom of religion.
c. schools.
d. more freedom for women.

____ **5.** The five Central Asian republics became independent when the Soviet Union broke up in
a. 1975.
b. 1986.
c. 1991.
d. 1999.

____ **6.** All of the Central Asian countries are switching from the Cyrillic alphabet to the
a. Russian alphabet.
b. English alphabet.
c. Greek alphabet.
d. Latin alphabet.

____ **7.** All five of the Central Asian countries are now ruled by
a. Soviet-style governments.
b. British-style governments.
c. American-style governments.
d. Japanese-style governments.

____ **8.** All of the Central Asian countries have strong economic ties to
a. the United States.
b. Russia.
c. Germany.
d. Great Britain.

____ **9.** A major crop in Central Asia is
a. henequen.
b. flax.
c. corn.
d. cotton.

____ **10.** People who often move from place to place are called
a. caravans.
b. Arals.
c. nomads.
d. Silk Roaders.

Central Asia

CHAPTER 19

SECTON 3

Fill in the Blank • *(10 points each)* For each of the following statements, fill in the blank with the appropriate word, phrase, or name.

1. Of the Central Asian states, Kazakhstan was the first to be conquered by

   _____, and that country's influence is still strong there.

2. Kazakhs celebrate the _____ twice, once on January 1 and again on Nauruz.

3. _____ has many mountains, and the people live mostly in the valleys there.

4. The _____, a movable round house of wool felt mats over a wood frame, is important to the Kyrgyz.

5. _____ membership is still important in Kyrgyzstan.

6. In 1993 Turkmenistan adopted _____, rather than Russian, as the country's second official language.

7. _____ has the largest population of all the Central Asian countries.

8. People who live in Uzbekistan are required to study the Uzbek language in order

   to be eligible for _____.

9. In the mid-1990s Tajikistan experienced a(n) _____.

10. Tajiks consider the great literature written in _____ to be part of their cultural heritage.

The Arabian Peninsula, Iraq, Iran, and Afghanistan

CHAPTER 20

SECTION 1

Matching • *(10 points each)* Match each of the following terms or places with the correct description by writing the letter of the description in the space provided. Some descriptions will not be used.

_____ **1.** Arabian Peninsula

_____ **2.** Mesopotamia

_____ **3.** exotic rivers

_____ **4.** Iran

_____ **5.** Yemen

_____ **6.** Hindu Kush

_____ **7.** Saudi Arabia

_____ **8.** wadis

_____ **9.** fossil water

_____ **10.** oil

a. Country that is bordered by the Elburz Mountains and Kopet-Dag range in the north and the Zagros Mountains in the southwest

b. Gulf whose shores are the location of most of the region's oilfields

c. Ancient name given to the area between the Tigris and Euphrates Rivers

d. Water that is not being replaced by rainfall

e. Country in which the Arabian Peninsula reaches its highest point

f. Term for dry streambeds

g. Large rectangular area bordered by the Red Sea, Gulf of Aden, Arabian Sea, and Persian Gulf

h. Country that has the largest sand desert in the world

i. The region's most important resource

j. Rivers that begin in humid regions and then flow through dry areas

k. Valuable resource found in the Persian Gulf oyster beds

l. Mountain range in Afghanistan

The Arabian Peninsula, Iraq, Iran, and Afghanistan

CHAPTER 20

SECTION 2

Multiple Choice • *(10 points each)* For each of the following, write the letter of the *best* choice in the space provided.

_____ **1.** The largest country of the Arabian Peninsula is
   a. Kuwait.
   b. Bahrain.
   c. Saudi Arabia.
   d. Qatar.

_____ **2.** Muhammad was the founder of
   a. Islam.
   b. OPEC.
   c. the Arabic language.
   d. the United Arab Emirates.

_____ **3.** In 1990 Kuwait was invaded by
   a. Iran.
   b. Saudi Arabia.
   c. the United Arab Emirates.
   d. Iraq.

_____ **4.** The capital of Saudi Arabia is
   a. Tehran.
   b. Riyadh.
   c. Sunni.
   d. Yemen.

_____ **5.** Which of the following statements is incorrect?
   a. Saudi women rarely appear in public without a husband or male relative.
   b. Saudi women do not drive cars.
   c. Saudi women make up more than 50 percent of the Saudi workforce.
   d. The freedom of Saudi women is limited by Saudi law.

_____ **6.** The most important part of the Saudi economy involves
   a. automobiles.
   b. oil.
   c. precious gems.
   d. gold and silver.

_____ **7.** Which of the following countries is not a monarchy?
   a. Yemen
   b. Oman
   c. Bahrain
   d. Qatar

_____ **8.** The Qur'an is the holy book of which group of people?
   a. Jews
   b. Protestants
   c. Catholics
   d. Muslims

_____ **9.** The economy of the United Arab Emirates is heavily dependent on
   a. oil.
   b. diamonds.
   c. corn.
   d. cotton.

_____ **10.** The poorest country on the Arabian Peninsula is
   a. Oman.
   b. Yemen.
   c. Qatar.
   d. Bahrain.

The Arabian Peninsula, Iraq, Iran, and Afghanistan

CHAPTER 20

SECTION 3

Fill in the Blank • *(10 points each)* For each of the following statements, fill in the blank with the appropriate word, phrase, or name.

1. In the A.D. 600s the _____ conquered Mesopotamia.

2. In the 1500s Mesopotamia became part of the _____ Empire.

3. _____ became president of Iraq and leader of the Iraqi armed forces in the late 1970s.

4. In 1980 Iraq invaded _____, leading to a war that continued until 1988.

5. The _____ War was fought in 1991 to force Iraqi invaders to leave the country of Kuwait.

6. Iraq has the world's second-largest known reserve of _____.

7. Several years ago the United Nations placed a(n) _____, or limit on trade, on Iraq.

8. Many of Iraq's factories produce _____ for the country's large army.

9. Iraq's farming sector is supported by irrigation from the _____ and Euphrates Rivers.

10. Nearly all Iraqis are _____, or followers of the Islamic faith.

The Arabian Peninsula, Iraq, Iran, and Afghanistan

CHAPTER 20

SECTION 4

True/False • *(10 points each)* Indicate whether each statement below is true or false by writing *T* or *F* in the space provided. If the statement is false, use the space below to explain why.

_____ **1.** Alexander the Great conquered the Persian Empire in the 300s B.C.

_____ **2.** In 1942 an Iranian military officer took power in Iran under the old Persian title of chief.

_____ **3.** In 1979 Iranian students attacked the U.S. embassy in Tehran and took British officials hostage for more than a year.

_____ **4.** Iran is a theocracy, or a government ruled by religious leaders.

_____ **5.** Automobile manufacturing is Iran's main industry.

_____ **6.** About one third of Iran's workforce is employed in agriculture.

_____ **7.** The official language of Iran is Persian.

_____ **8.** Afghanistan is a landlocked country that contains high mountains and fertile valleys.

_____ **9.** After the Soviet Union sent troops into Afghanistan in 1979, there was a long conflict between Soviet troops and Afghan rebels.

_____ **10.** The Taliban have encouraged women in Afghanistan to work outside the home.

The Eastern Mediterranean

CHAPTER 21

SECTION 1

Fill in the Blank • *(10 points each)* For each of the following statements, fill in the blank with the appropriate word, phrase, or name.

1. The country of Israel controls an area known as the _____, which includes the West Bank, Gaza Strip, and Golan Heights.

2. The Dardanelles, the Bosporus, and the Sea of Marmara separate Europe from

   _____.

3. The two main ridges in the region are separated by the _____ River valley.

4. The _____ is so salty that swimmers cannot sink in it.

5. Turkey's Black Sea coast and the Mediterranean coast all the way to Israel have

   a(n) _____ type of climate.

6. A(n) _____ climate can be found in central Syria and lands farther south.

7. The _____ Desert covers much of Syria and Jordan.

8. A desert called the _____ lies in the southern part of Israel.

9. Unlike some nearby countries, the countries of the eastern Mediterranean do not

   have large deposits of _____.

10. The Dead Sea is a source of _____.

The Eastern Mediterranean

SECTION 2

Matching • *(10 points each)* Match each of the following terms, places, or names with the correct description by writing the letter of the description in the space provided. Some descriptions will not be used.

_____ **1.** Byzantium

_____ **2.** Seljuk Turks

_____ **3.** Ottoman Turks

_____ **4.** Alexander the Great

_____ **5.** Mustafa Kemal

_____ **6.** Ankara

_____ **7.** fez

_____ **8.** secular

_____ **9.** Tigris and Euphrates Rivers

_____ **10.** shish kebab

a. Person who dissolved the Ottoman Empire and created the nation of Turkey

b. Grilled meat on a skewer, a favorite Turkish dish

c. Important city that was later renamed Constantinople

d. Traditional Turkish hat banned by Kemal Atatürk

e. Largest minority group in Turkey

f. Legislature of Turkey

g. Nomadic people from Central Asia who invaded Asia Minor in the A.D. 1000s

h. Site where Turkey began building large dams in the 1990s

i. Person who conquered Asia Minor in the 330s B.C.

j. Capital of Turkey

k. Term that means that religion is kept out of a country's government

l. Turkish people who conquered Constantinople in 1453

The Eastern Mediterranean

SECTION 3

True/False • *(10 points each)* Indicate whether each statement below is true or false by writing *T* or *F* in the space provided. If the statement is false, use the space below to explain why.

_____ **1.** The Catholics first established the kingdom of Israel around 3000 years ago.

_____ **2.** The forced scattering of the Jews by the Romans is known as Zionism.

_____ **3.** Both Roman and Jewish rulers viewed the teachings of Jesus as dangerous.

_____ **4.** The Crusades were a series of invasions launched by Europeans from the 1000s to the 1200s to take Jerusalem from the Arabs.

_____ **5.** At the end of World War I, Palestine came under the control of the United States.

_____ **6.** The State of Israel was created in 1948 by the Jews in Palestine.

_____ **7.** Israel's parliament is known as the Diaspora.

_____ **8.** Both Hebrew and Arabic are Israel's official languages.

_____ **9.** The Gaza Strip is the largest of the occupied areas.

_____ **10.** Control of Jerusalem is an emotional issue because the city contains sites that are holy to Jews, Muslims, and Christians.

The Eastern Mediterranean

SECTION 4

Multiple Choice • *(10 points each)* For each of the following, write the letter of the *best* choice in the space provided.

_____ **1.** After the creation of Israel and the war of 1948, Jordan annexed the Arab lands of
a. Syria.
b. the West Bank.
c. the Golan Heights.
d. Lebanon.

_____ **2.** Until 1949 Jordan was known as
a. the Gaza Strip.
b. Amman.
c. Transjordan.
d. Palestine.

_____ **3.** Which country gained control of Lebanon after World War I?
a. Israel
b. France
c. the United States
d. Great Britain

_____ **4.** About 90 percent of Syria's population is made up of
a. Arabs.
b. Christians.
c. Kurds.
d. Armenians.

_____ **5.** Which of the following countries drew the borders of Jordan?
a. Germany
b. Great Britain
c. Italy
d. France

_____ **6.** Syria's most important manufactured products are
a. textiles.
b. automobiles.
c. jewelry items.
d. weapons.

_____ **7.** The majority of people in Lebanon are either Muslim or
a. Buddhist.
b. Hindu.
c. Jewish.
d. Christian.

_____ **8.** Syria became independent from France in
a. the 1820s.
b. the 1910s.
c. the 1940s.
d. the 1990s.

_____ **9.** How do most people in Lebanon identify themselves?
a. by their country of origin
b. by their race
c. by their occupation
d. by their religion

_____ **10.** The capital of Syria is
a. Beirut.
b. Damascus.
c. Jordan.
d. Amman.

North Africa

True/False • *(10 points each)* Indicate whether each statement below is true or false by writing *T* or *F* in the space provided. If the statement is false, use the space below to explain why.

_____ **1.** North Africa stretches from the Pacific Ocean to the Red Sea.

_____ **2.** The name *Sahara* comes from the Arabic word for "mountain."

_____ **3.** Great "seas" of sand dunes called ergs cover about a quarter of the Sahara.

_____ **4.** The Sahara contains both mountains and very low areas, called depressions.

_____ **5.** The world's longest river, the Nile, is formed by the union of two rivers—the Blue Nile and the White Nile.

_____ **6.** The silt left by the Nile destroyed the soil used by farmers to grow crops.

_____ **7.** A steppe climate covers most of North Africa.

_____ **8.** No plants or animals can live in the Sahara.

_____ **9.** North Africa has good fishing waters.

_____ **10.** The Sahara contains mineral resources such as copper, gold, and silver.

CHAPTER 22 North Africa

Multiple Choice • *(10 points each)* For each of the following, write the letter of the *best* choice in the space provided.

_____ **1.** Pharaohs were Egyptian
   a. kings.
   b. pictures and symbols.
   c. slaves.
   d. buildings.

_____ **2.** What Egyptian city did Alexander the Great found in 332 B.C.?
   a. Cairo
   b. Fès
   c. Western Sahara
   d. Alexandria

_____ **3.** Today most North Africans speak Arabic and are
   a. Catholic.
   b. Jewish.
   c. Muslim.
   d. Buddhist.

_____ **4.** Libya, Morocco, and Tunisia won independence from European control in the
   a. 1890s.
   b. 1910s.
   c. 1950s.
   d. 1990s.

_____ **5.** Muslims abstain from food and drink during the day
   a. on their birthdays.
   b. during Ramadan.
   c. on the fourth of July.
   d. on New Year's day.

_____ **6.** In 1979 Egypt signed a peace treaty with
   a. Israel.
   b. the United States.
   c. Morocco.
   d. Canada.

_____ **7.** Which of the following countries has the largest number of non-Muslims?
   a. Egypt
   b. Libya
   c. Tunisia
   d. Morocco

_____ **8.** The food known as couscous looks like small pellets of
   a. broccoli.
   b. salt and pepper.
   c. corn.
   d. pasta.

_____ **9.** The last North African country to gain independence was
   a. Morocco.
   b. Algeria.
   c. Tunisia.
   d. Libya.

_____ **10.** A *sintir* is a(n)
   a. holy person.
   b. Moroccan food.
   c. musical instrument.
   d. Muslim holiday.

North Africa

Fill in the Blank • *(10 points each)* For each of the following statements, fill in the blank with the appropriate word, phrase, or name.

1. North Africa's most populous country is _____.

2. Most rural Egyptians are farmers, known as _____, who own very small plots of land.

3. Egypt's largest city is _____, which is also its capital.

4. Cairo lies along old trading routes between Asia and _____.

5. Alexandria, Egypt's second-largest city, is located in the Nile Delta along the

   _____ coast.

6. Three of the most important industries in Egypt are _____, tourism, and oil.

7. The _____ Canal is an important source of income for Egypt because ships must pay tolls to pass through it.

8. About 40 percent of Egyptian workers are _____.

9. _____ of the soil in Egypt has brought to the surface salts that are harmful to crops.

10. One challenge Egyptians face is the debate over the role of _____ in the country.

North Africa

SECTION 4

Matching • *(10 points each)* Match each of the following terms, places, or names with the correct description by writing the letter of the description in the space provided. Some descriptions will not be used.

_____ **1.** Maghreb   **a.** Capital of Algeria

_____ **2.** Libya   **b.** Algerian marketplaces

_____ **3.** Tripoli   **c.** Most urbanized country in the region

_____ **4.** Algiers   **d.** Only North African country that has little oil

_____ **5.** Casbah   **e.** Largest city in the region

_____ **6.** free port   **f.** Name used for western Libya, Tunisia, Algeria, and Morocco

_____ **7.** Morocco   **g.** Moroccan city that overlooks the Strait of Gibraltar

_____ **8.** dictator   **h.** Old district of Algiers

_____ **9.** souks   **i.** Someone who rules a country with complete power

_____ **10.** Cairo   **j.** City in which almost no taxes are placed on goods sold there

     k. Capital of Libya

     l. Most important resource in North Africa

West Africa

SECTION 1

Fill in the Blank • *(10 points each)* For each of the following statements, fill in the blank with the appropriate word, phrase, or name.

1. West Africa's climates are _____, meaning that they stretch from east to west in bands or zones.

2. The _____ is the world's largest desert.

3. South of the Sahara is a region of dry grasslands called the _____.

4. During the winter season, a dry, dusty wind known as the _____ blows south from the Sahara.

5. South of the Sahel is the _____ zone, which contains good soil, thick grass, and scattered tall trees.

6. The _____ fly carries sleeping sickness, a disease that can kill animals and humans.

7. Many of West Africa's largest cities can be found in the _____ zone.

8. The _____ River is the most important river in West Africa.

9. West Africa's mineral riches include _____, the main source of aluminum.

10. _____ and related products make up about 95 percent of Nigeria's exports.

West Africa

CHAPTER 23

Matching • *(10 points each)* Match each of the following terms or places with the correct description by writing the letter of the description in the space provided. Some descriptions will not be used.

_____ **1.** archaeology

_____ **2.** oral history

_____ **3.** Ghana

_____ **4.** Mali

_____ **5.** Songhay

_____ **6.** Gold Coast

_____ **7.** Brazil

_____ **8.** France

_____ **9.** Liberia

_____ **10.** Portugal

a. Country that claimed most of the West African region in the 1800s

b. Study of the remains and ruins of past cultures

c. Last European country to give up its West African colonies, in 1974

d. Name given to the west coast of Africa by Europeans

e. One of the earliest West African kingdoms, which was based on trade

f. Main religion of the Sahel

g. Belief that bodies of water, animals, trees, and other natural objects have spirits

h. Spoken information passed down through generations from person to person

i. Along with the West Indies, place to which the greatest number of West African slaves were taken

j. West African kingdom whose city of Timbuktu was a major cultural center

k. Only West African country to remain independent in the 1800s

l. West African kingdom whose king was Mansa Mūsā, known for his wealth and wise rule

West Africa

True/False • *(10 points each)* Indicate whether each statement below is true or false by writing *T* or *F* in the space provided. If the statement is false, use the space below to explain why.

_____ **1.** Most of the people who live in Mauritania, Mali, and Niger are Christian.

_____ **2.** The official language in Mali and Niger is French.

_____ **3.** In the savanna regions farmers grow millet and sorghum because those grains can survive drought.

_____ **4.** Many Mauritanians are Moors, people of mixed Arab and Berber origin.

_____ **5.** The country of Mali is surrounded by water on three sides.

_____ **6.** Nearly 100 percent of Niger's land is good for farming.

_____ **7.** Desert nomads in Niger depend on dairy products from their herds for food.

_____ **8.** Chad and Burkina Faso are among the world's richest and most developed countries.

_____ **9.** Today Lake Chad is only one third as large as it was in 1950.

_____ **10.** The name *Burkina Faso* means "land of the honest people."

West Africa

CHAPTER 23

SECTION 4

Multiple Choice • *(10 points each)* For each of the following, write the letter of the *best* choice in the space provided.

____ 1. The largest country along West Africa's coast is
   a. Gambia.
   b. Senegal.
   c. Nigeria.
   d. Liberia.

____ 2. Which of the following is West Africa's only island country?
   a. Sierra Leone
   b. Ghana
   c. Cape Verde
   d. Senegal

____ 3. The largest Christian church building in Africa can be found in
   a. Liberia.
   b. Côte d'Ivoire.
   c. Benin.
   d. Gambia.

____ 4. Côte d'Ivoire is a former French colony whose name in English means
   a. "Coat of Armor."
   b. "Ivory Ghost."
   c. "Nest of Gold."
   d. "Ivory Coast."

____ 5. Liberia was settled by
   a. French pirates.
   b. British criminals.
   c. Jewish immigrants.
   d. freed American slaves.

____ 6. Which of the following groups tried to secede from Nigeria in the 1960s?
   a. Ibo
   b. Yoruba
   c. Fula
   d. Hausa

____ 7. Guinea's main resource is
   a. silver.
   b. bauxite.
   c. oil.
   d. iron ore.

____ 8. Which country contains one of the largest human-made lakes in the world?
   a. Ghana
   b. Togo
   c. Senegal
   d. Nigeria

____ 9. The most important natural resource in Nigeria is
   a. natural gas.
   b. diamonds.
   c. gold.
   d. oil.

____ 10. What is the most important crop in Senegal and Gambia?
   a. cotton
   b. peanuts
   c. tomatoes
   d. potatoes

Name _____ Class _____ Date _____

East Africa

True/False • *(10 points each)* Indicate whether each statement below is true or false by writing *T* or *F* in the space provided. If the statement is false, use the space below to explain why.

_____ **1.** East Africa is a land of mostly deserts.

_____ **2.** East Africa's most striking features are its great rifts.

_____ **3.** Mount Kilimanjaro is Africa's shortest mountain.

_____ **4.** Plains along the eastern rift in Tanzania and Kenya are home to well-known national parks.

_____ **5.** The Nile, the world's longest river, begins in East Africa and flows north to the Mediterranean Sea.

_____ **6.** Lake Victoria is the source of the Blue Nile.

_____ **7.** The White Nile and the Blue Nile meet at Khartoum, where they form the Nile.

_____ **8.** Lake Victoria is Africa's largest lake in area, but it is deep.

_____ **9.** Lake Nakuru is too salty for most fish to live in it.

_____ **10.** Most East Africans are farmers or herders.

East Africa

SECTION 2

Matching • *(10 points each)* Match each of the following terms or places with the correct description by writing the letter of the description in the space provided. Some descriptions will not be used.

_____ **1.** Meroë

_____ **2.** Nubia

_____ **3.** Islam

_____ **4.** Christianity

_____ **5.** Portugal

_____ **6.** Zanzibar

_____ **7.** Great Britain

_____ **8.** Kenya

_____ **9.** Ethiopia

_____ **10.** Swahili

a. East African island that became an international slave-trading center in the late 1700s

b. Only East African country that was never colonized by Europeans, although it once was annexed by Italy

c. Religion believed to have been introduced in Ethiopia as early as the A.D. 300s

d. Only East African colony where large numbers of Europeans settled

e. Country that set up the first European forts and settlements on the East African coast

f. Country that has been engaged in ethnic conflict with Burundi

g. Place where the branches of the Nile come together and where several early civilizations developed

h. Country that has been involved in ethnic conflict with Rwanda

i. Religion brought to northern Sudan by Arabic-speaking nomads from Egypt

j. Bantu language that has been greatly influenced by Arabic

k. Country that gained control over much of East Africa in the 1880s

l. Christianity spread here in the A.D. 500s

East Africa

SECTION 3

Multiple Choice • *(10 points each)* For each of the following, write the letter of the *best* choice in the space provided.

_____ **1.** Kenya's first cities were founded along the coast of the
   a. Atlantic Ocean.
   b. Pacific Ocean.
   c. Mediterranean Sea.
   d. Indian Ocean.

_____ **2.** Africa's largest country is
   a. Tanzania.
   b. Uganda.
   c. Rwanda.
   d. Sudan.

_____ **3.** Kenya gained independence from Great Britain in the
   a. 1890s.
   b. 1920s.
   c. 1960s.
   d. 1990s.

_____ **4.** Tanzania was created by the uniting of Tanganyika and
   a. Kenya.
   b. Zanzibar.
   c. Zaire.
   d. Burundi.

_____ **5.** Where has evidence of some of the earliest humanlike existence been found?
   a. Olduvai Gorge
   b. Mount Kilimanjaro
   c. Serengeti Plain
   d. Rwanda

_____ **6.** Rwanda and Burundi were once colonies of which European country?
   a. Italy
   b. Spain
   c. Great Britain
   d. Germany

_____ **7.** Which of the following countries lies north and west of Lake Victoria?
   a. Uganda
   b. Sudan
   c. Burundi
   d. Kenya

_____ **8.** When did the economy of Uganda collapse?
   a. 1910s
   b. 1940s
   c. 1970s
   d. 1990s

_____ **9.** Beginning in the 1500s the coast of the Indian Ocean was controlled by
   a. France.
   b. Portugal.
   c. the United States.
   d. Spain.

_____ **10.** Khartoum is the capital of
   a. Zanzibar.
   b. Sudan.
   c. Burundi.
   d. Kenya.

East Africa

SECTION 4

Fill in the Blank • *(10 points each)* For each of the following statements, fill in the blank with the appropriate word, phrase, or name.

1. Several million Ethiopians starved in the _____.

2. Although it exports coffee, livestock, and oilseeds, _____ is one of the world's poorest countries.

3. Periods of little rain that damage crops are known as _____.

4. Most of Ethiopia's lowland people are _____.

5. Eritrea is located on the _____ Sea.

6. In the late 1800s the European country of _____ made the area of Eritrea a colony.

7. _____, troubled by civil war and drought, received aid from the United Nations.

8. Most Somali share the same religion, which is _____.

9. _____ is a small desert country located on the Bab al-Mandab.

10. The people of Djibouti include the _____ and the Afar.

Central Africa

True/False • *(10 points each)* Indicate whether each statement below is true or false by writing *T* or *F* in the space provided. If the statement is false, use the space below to explain why.

_____ **1.** Central Africa stretches southward from Cameroon and the Central African Republic to Angola and Zambia.

_____ **2.** Volcanic mountains can be found in northwestern Cameroon.

_____ **3.** In the northern part of the region, the Congo River flows westward to the Pacific Ocean.

_____ **4.** In the southern part of the region, the Zambezi River flows eastward to the Indian Ocean.

_____ **5.** The Congo River is famous for its great falls and hydroelectric dams and lakes.

_____ **6.** Central Africa lies along the equator and in the low latitudes.

_____ **7.** The many different kinds of animals in the tropical rain forest form a complete canopy.

_____ **8.** The rapid clearing of the tropical rain forest threatens the plants, animals, and people who live there.

_____ **9.** Central Africa's rivers are among its most important natural resources.

_____ **10.** Most of the copper in Africa can be found in an area known as the metal belt.

Central Africa

SECTION 2

Multiple Choice • *(10 points each)* For each of the following, write the letter of the *best* choice in the space provided.

_____ **1.** The most common language in central Africa today is
  a. English.
  b. Bantu.
  c. French.
  d. German.

_____ **2.** The Kongo Kingdom was located at the mouth of
  a. the Congo River.
  b. the Western Rift Valley.
  c. Lake Tanganyika.
  d. the Zambezi River.

_____ **3.** African colonies did not gain independence until after
  a. World War I.
  b. World War II.
  c. the Korean War.
  d. the Vietnam War.

_____ **4.** What was the last European colony in central Africa?
  a. Belgian Congo
  b. Gabon
  c. Cameroon
  d. Angola

_____ **5.** Where is English the official language?
  a. Equatorial Guinea
  b. Angola
  c. Zambia and Malawi
  d. Democratic Republic of the Congo

_____ **6.** Many people in the former French, Spanish, and Portuguese colonies are
  a. Jewish.
  b. Protestant Christian.
  c. Muslim.
  d. Roman Catholic.

_____ **7.** What is *fufu?*
  a. a food
  b. an animal
  c. an article of clothing
  d. a musical instrument

_____ **8.** *Makossa* is a type of
  a. hair style.
  b. music.
  c. sculpture.
  d. food.

_____ **9.** About 2,000 years ago new peoples began to move into central Africa from
  a. eastern Africa.
  b. northern Africa.
  c. western Africa.
  d. southern Africa.

_____ **10.** Europeans first arrived in central Africa in the late
  a. 1200s.
  b. 1400s.
  c. 1600s.
  d. 1800s.

Central Africa

CHAPTER 25

Fill in the Blank • *(10 points each)* For each of the following statements, fill in the blank with the appropriate word, phrase, or name.

1. _____ sailors first made contact with the Kongo Kingdom in 1482.

2. In the 1870s King Leopold II of _____ took control of the Congo Basin.

3. The Belgian businesses and people who moved to the Congo Free State mined

   _____ and other resources.

4. The Congo Free State won independence from Belgium in _____.

5. Mobutu Sese Seko changed the name of the Congo Free State to _____ in 1971.

6. A(n) _____ is a war between two or more groups within a country.

7. The _____ people are among the largest ethnic groups in the Democratic Republic of the Congo today.

8. The official language in the Democratic Republic of the Congo today is

   _____.

9. _____ is the capital and largest city in the Democratic Republic of the Congo.

10. The southern section of the Democratic Republic of the Congo is part of central

    Africa's rich _____ belt.

Central Africa

SECTION 4

Matching • *(10 points each)* Match each of the following places with the correct description by writing the letter of the description in the space provided. Some descriptions will not be used.

_____ **1.** Cameroon

_____ **2.** Spain

_____ **3.** Central African Republic

_____ **4.** Yaoundé

_____ **5.** Douala

_____ **6.** Brazzaville

_____ **7.** Gabon

_____ **8.** Great Britain

_____ **9.** Portugal

_____ **10.** Luanda

a. Capital of Cameroon

b. Country that has the strongest economy in central Africa

c. Country that granted independence to Angola in 1975

d. Important seaport on the Atlantic coast

e. Part of a country separated by territory of other countries

f. Country that gave Equatorial Guinea independence in 1968

g. Capital of Angola

h. Second-most-populous country in central Africa

i. Capital of the Republic of the Congo

j. Country whose rich copper mines provide much of its income

k. Country that granted independence to Zambia and Malawi in 1964

l. Country that had been a German colony until after World War I

Southern Africa

True/False • *(10 points each)* Indicate whether each statement below is true or false by writing *T* or *F* in the space provided. If the statement is false, use the space below to explain why.

_____ **1.** Madagascar is the world's fourth-largest island.

_____ **2.** The southeastern edge of southern Africa's interior plateau is a mountain range called the Drakensberg.

_____ **3.** The open grassland areas of South Africa are called enclaves.

_____ **4.** Kruger National Park is designed to protect one species of animal, the lion.

_____ **5.** Winds carry moisture to southern Africa from the Pacific Ocean.

_____ **6.** Near the Cape of Good Hope, winter rains and summer drought create a Mediterranean climate.

_____ **7.** The Kalahari and the Namib are two major mountains in southern Africa.

_____ **8.** Southern Africa has some of the world's most spectacular rivers and waterfalls.

_____ **9.** The Orange and the Limpopo are two of the region's major rivers.

_____ **10.** Southern Africa is very poor in mineral resources.

Southern Africa

CHAPTER 26

Matching • *(10 points each)* Match each of the following terms or places with the correct description by writing the letter of the description in the space provided. Some descriptions will not be used.

_____ **1.** Khoisan

_____ **2.** Bantu

_____ **3.** *zimbabwe*

_____ **4.** porcelain

_____ **5.** Madagascar

_____ **6.** Mozambique

_____ **7.** Afrikaners

_____ **8.** Boers

_____ **9.** diamonds

_____ **10.** gold

a. Country placed under South Africa's control after Germany's defeat in World War I

b. Family of languages that share unusual "click" sounds

c. Country in which the Portuguese set up forts in the early 1500s

d. Animal completely wiped out by hunters in some parts of southern Africa

e. Chinese pottery found at Great Zimbabwe, suggesting that Africa and East Asia were connected by an Indian Ocean trade network

f. Resource discovered in the Transvaal in 1886

g. Family of languages that spread from central Africa into southern Africa between about 2,000 and 1,500 years ago

h. Resource found in the northern Cape Colony in the 1860s

i. White descendants of the original colonists in South Africa

j. Stone-walled towns built by the Shona

k. Afrikaner frontier farmers

l. Island whose people blend cultural features of Africa and Asia

Southern Africa

SECTION 3

Multiple Choice • *(10 points each)* For each of the following, write the letter of the *best* choice in the space provided.

_____ **1.** South Africa's policy of apartheid
    a. ended trade with Western nations.
    b. separated its different peoples.
    c. made Islam its official religion.
    d. provided free education for all.

_____ **2.** People who protested apartheid laws
    a. were sent to prison.
    b. were appointed to government.
    c. were ignored.
    d. were given the better jobs.

_____ **3.** Blacks who worked in white-owned businesses had to live in special areas called
    a. sanctions.
    b. townships.
    c. cities.
    d. communities.

_____ **4.** In 1994 Nelson Mandela
    a. was sent to prison.
    b. moved to the United States.
    c. was elected president of South Africa.
    d. was killed by Afrikaners.

_____ **5.** Zimbabwe once was known as
    a. Northern Rhodesia.
    b. Western Rhodesia.
    c. Southern Rhodesia.
    d. Eastern Rhodesia.

_____ **6.** How did many people around the world react to apartheid laws?
    a. They did not know about the laws.
    b. They ignored the laws.
    c. They supported the laws.
    d. They objected to the laws.

_____ **7.** Today, all races in South Africa
    a. have equal rights.
    b. have equal amounts of wealth.
    c. are separated by law.
    d. are prohibited from voting.

_____ **8.** South Africa's energy resources include
    a. coal and oil.
    b. coal and hydroelectric power.
    c. oil and natural gas.
    d. natural gas and hydroelectric power.

_____ **9.** South Africa's major port is
    a. Johannesburg.
    b. Cape Town.
    c. Port Elizabeth.
    d. Durban.

_____ **10.** How many official languages does South Africa have?
    a. 1
    b. 11
    c. 21
    d. 31

Southern Africa

Fill in the Blank • *(10 points each)* For each of the following statements, fill in the blank with the appropriate word, phrase, or name.

1. The capital of _____, Windhoek, is located in the highlands.

2. The official language of Namibia is _____.

3. _____ is a large, landlocked, semiarid country.

4. About 79 percent of the people in Botswana belong to one ethnic group, known as

   the _____.

5. The capital of Zimbabwe, Harare, was called _____ when the country was known as Southern Rhodesia.

6. Although _____ in Zimbabwe make up less than 1 percent of the population, they still own most of the large farms and ranches.

7. The economy of _____ was badly damaged by civil war, and so it is one of the world's poorest countries.

8. Most of the people in Mozambique belong to various _____ ethnic groups.

9. Madagascar was once a colony of the European country of _____.

10. Most of the people who live in Madagascar depend on _____ farming.

Name _____ Class _____ Date _____

China, Mongolia, and Taiwan

CHAPTER 27

Matching • *(10 points each)* Match each of the following terms or places with the correct description by writing the letter of the description in the space provided. Some descriptions will not be used.

_____ **1.** China

_____ **2.** Taiwan

_____ **3.** Himalayas

_____ **4.** Mount Everest

_____ **5.** Plateau of Tibet

_____ **6.** Gobi

_____ **7.** North China Plain

_____ **8.** Chang

_____ **9.** dikes

_____ **10.** typhoons

a. World's tallest mountain

b. River whose name means "yellow river"

c. Country that has the world's largest population

d. Violent storms with high winds and heavy rains

e. Coldest desert in the world

f. High banks of earth or concrete used to help reduce flooding

g. China's—and Asia's—longest river

h. Tropical island just off mainland China's coast

i. Largest plain in China

j. Tallest mountain range in the world

k. Southern China's most important river and transportation route

l. World's highest plateau

China, Mongolia, and Taiwan

CHAPTER 27

SECTION 2

Multiple Choice • *(10 points each)* For each of the following, write the letter of the *best* choice in the space provided.

_____ 1. One of the region's main sources of food is
   a. corn.
   b. potatoes.
   c. squash.
   d. rice.

_____ 2. The People's Republic of China was led by
   a. Mao Zedong.
   b. Confucius.
   c. Sun Yat-sen.
   d. Genghis Khan.

_____ 3. During the Han dynasty, the Chinese invented
   a. radio.
   b. electricity.
   c. the computer.
   d. the compass

_____ 4. In the 1200s China was conquered by the
   a. French.
   b. Japanese.
   c. Mongols.
   d. Afrikaners.

_____ 5. Which country was the first to set up a trade colony in south China in the 1500s?
   a. Great Britain
   b. Portugal
   c. the United States
   d. Italy

_____ 6. After Sun Yat-sen's death, Chinese revolutionaries split into two groups, the
   a. Democrats and the Republicans.
   b. Socialists and the Communists.
   c. Nationalists and the Democrats.
   d. Nationalists and the Communists.

_____ 7. The emperors of which dynasty ordered the building of the Great Wall?
   a. Qin
   b. Han
   c. Ming
   d. Qing

_____ 8. The Republic of China was led by
   a. Sun Yat-sen.
   b. Chiang Kai-shek.
   c. Genghis Khan.
   d. Mao Zedong.

_____ 9. Siddhartha Gautama founded the religion of
   a. Hinduism.
   b. Daoism.
   c. Buddhism.
   d. Confucianism.

_____ 10. How many years of school are Chinese children required to attend?
   a. 5
   b. 7
   c. 9
   d. 12

China, Mongolia, and Taiwan

SECTION 3

Fill in the Blank • *(10 points each)* For each of the following statements, fill in the blank with the appropriate word, phrase, or name.

1. More people live in _____ than in Europe, Russia, and the United States combined.

2. The _____ half of China is almost empty of people.

3. China has _____ cities with populations larger than 1 million.

4. _____, also known as Peking, is the capital of China.

5. China's largest city, _____, lies on the Chang Delta.

6. With a population of 6.5 million, _____, a former British colony, is one of the world's most densely populated places.

7. Macao, a former _____ colony, was the last foreign territory in China.

8. Only about _____ percent of the land in China is suitable for farming.

9. China is a leading producer of _____ ore.

10. China's _____ status means that it gets special trade advantages from the United States.

China, Mongolia, and Taiwan

SECTION 4

True/False • *(10 points each)* Indicate whether each statement below is true or false by writing *T* or *F* in the space provided. If the statement is false, use the space below to explain why.

_____ **1.** Led by Genghis Khan, the Mongols conquered most of Asia, including China.

_____ **2.** The Mongol Empire reached its height in the late 1900s.

_____ **3.** In 1911, with Russian support, Mongolia declared its independence from China.

_____ **4.** After Mongolia declared its independence, it came under the influence of the United States.

_____ **5.** Ulaanbaatar is Mongolia's capital city and also its main industrial and commercial center.

_____ **6.** For many years Taiwan was known in the West as Formosa.

_____ **7.** The search for gold and silver brought Europeans to Taiwan.

_____ **8.** After Japan's surrender at the end of World War II, Germany took control of Taiwan.

_____ **9.** Chinese ways dominate the culture of Taiwan today.

_____ **10.** The chief food crop in Taiwan today is rice.

Japan and the Koreas

CHAPTER 28

Multiple Choice • *(10 points each)* For each of the following, write the letter of the *best* choice in the space provided.

_____ **1.** The Korean Peninsula extends south-ward about 600 miles (965 km) from
   a. southern Japan.
   b. mainland Asia.
   c. Alaska.
   d. Australia.

_____ **2.** Japan's four home islands are
   a. Ryukyu, Okinawa, Kyushu, and Honshu.
   b. Fuji, Okinawa, Honshu, and Hokkaido.
   c. Hokkaido, Honshu, Shikoku, and Kyushu.
   d. Shinkansen, Yalu, Tumen, and Korea.

_____ **3.** Rugged mountains covered with forests are a common sight in
   a. South Korea.
   b. North Korea.
   c. Japan.
   d. all of the countries of this region.

_____ **4.** The two climate regions of Japan and the Koreas are
   a. subarctic and tundra climates.
   b. subarctic and desert climates.
   c. Mediterranean and steppe climates.
   d. humid continental and humid sub-tropical climates.

_____ **5.** What are tsunamis?
   a. large mountain animals
   b. minor earthquakes
   c. huge waves caused by undersea earthquakes
   d. large ocean storms similar to hurricanes

_____ **6.** What is the Pacific Ring of Fire?
   a. a zone where tectonic plates meet and where earthquakes are common
   b. a region of severe weather around the Pacific Ocean
   c. a wedding ceremony that is popu-lar in parts of Japan
   d. a military alliance between South Korea, Japan, and the United States

_____ **7.** The only area in the region that is rich in natural resources is
   a. North Korea.
   b. South Korea.
   c. Japan.
   d. Okinawa.

_____ **8.** North Korea contains
   a. oil and natural gas.
   b. iron ore, copper, zinc, lead, and coal.
   c. uranium, diamonds, and coal.
   d. rain forests and jungles.

_____ **9.** Many of the mountains in Japan were formed by
   a. asteroids.
   b. volcanic activity.
   c. huge landslides in prehistoric times.
   d. unknown events.

_____ **10.** The Oyashio and Japan Currents are
   a. wind patterns that affect eastern Asia.
   b. currents that influence the climates of Japan.
   c. mysterious electrical currents that affect parts of the Pacific Ocean.
   d. musical groups popular in Japan.

Japan and the Koreas

SECTION 2

Fill in the Blank • *(10 points each)* For each of the following statements, fill in the blank with the appropriate word, phrase, or place.

1. Japan's first inhabitants came from _____ thousands of years ago.

2. The earliest known religion of Japan, called _____, centers around the spirits of natural places, sacred animals, and ancestors.

3. _____, or priests, made the *kami's* wishes known to the Japanese people.

4. The principles of _____ include respect for elders, parents, and rulers.

5. Japanese lords hundreds of years ago were served by _____, or Japanese warriors.

6. The term _____ means "great general" and is the highest warrior rank.

7. Around 1900 Japan began to expand its empire as a way to obtain the

   _____ it needed to industrialize.

8. Japan annexed, or took control of, _____ in 1910.

9. The United States entered World War II in 1941 after Japan attacked the U.S. naval

   base at _____.

10. The Japanese government today is made up of an elected legislature called the

    _____ and a prime minister.

Japan and the Koreas

SECTION 3

Matching • *(10 points each)* Match each of the following terms or places with the correct description by writing the letter of the description in the space provided. Some descriptions will not be used.

_____ **1.** arable

_____ **2.** megalopolis

_____ **3.** Tokyo

_____ **4.** Osaka

_____ **5.** futon

_____ **6.** kimono

_____ **7.** intensive cultivation

_____ **8.** Kobe

_____ **9.** protectionism

_____ **10.** trade surplus

a. Japan's capital and center of government

b. Growing food on every bit of available land

c. Term that means "fit for growing crops"

d. Traditional robe that many Japanese people wear for special occasions such as weddings

e. Belief that work in itself is worthwhile

f. Densely populated urban area that often includes more than one city as well as the surrounding suburban areas

g. An important seaport in Japan

h. Cutting the hillside into a series of small, flat fields for planting

i. Lightweight cotton mattress on which Japanese people sleep

j. Use of trade barriers to protect a country's industries from foreign competition

k. Situation in which a nation exports more than it imports

l. Japanese city that has been a trading center for centuries

Japan and the Koreas

SECTION 4

Multiple Choice • *(10 points each)* For each of the following, write the letter of the *best* choice in the space provided.

____ 1. Korea's earliest inhabitants were
   a. warrior peoples from southern Asia.
   b. nomadic hunters from north and central Asia.
   c. farmers from southeastern Asia.
   d. invaders from North America.

____ 2. Korea's original religion, shamanism,
   a. recognized the existence of one god.
   b. recognized the spirits of natural places and ancestors.
   c. is not important today.
   d. has never been very popular among the people of the region.

____ 3. When did the Koryo dynasty arise?
   a. 1500 B.C.
   b. A.D. 600s
   c. A.D. 900s
   d. A.D. 1900s

____ 4. During Korea's golden age, which began in the A.D. 600s,
   a. Korea was known for its architecture, painting, ceramics, and jewelry.
   b. the Koreans conquered much of Asia.
   c. Korean colonists settled throughout much of the Pacific region.
   d. Koreans explored parts of western North and South America.

____ 5. The Hangul alphabet
   a. contained thousands of Chinese characters.
   b. was similar to the English alphabet.
   c. had only 24 symbols.
   d. is used throughout Asia today.

____ 6. During the period of the Hermit Kingdom,
   a. Koreans opened their country to people from around the world.
   b. Koreans conquered most of China.
   c. Korea was closed off to most other outsiders except China.
   d. Koreans threw off Chinese rule.

____ 7. When did Japan annex Korea?
   a. after World War II ended
   b. in the 1600s
   c. during World War I
   d. in 1910

____ 8. At the end of World War II, the Soviets
   a. helped communist leaders take power in North Korea.
   b. argued that Japan should keep control of all of Korea.
   c. helped the United States set up a new government in the Korean Peninsula.
   d. wanted to transfer control of Korea from Japan to China.

____ 9. North and South Korea are divided by a strip of land called
   a. the empty quarter.
   b. the uninhabited zone.
   c. the International Dateline.
   d. the demilitarized zone.

____ 10. During what years was the Korean War fought?
   a. 1941 to 1945
   b. 1950 to 1953
   c. 1954 to 1975
   d. 1898 to 1900

Japan and the Koreas

SECTION 5

True/False • *(10 points each)* Indicate whether each statement below is true or false by writing *T* or *F* in the space provided. If the statement is false, use the space below to explain why.

_____ **1.** South Korea's capital and largest city is Seoul.

_____ **2.** The rapid growth of South Korea's cities has brought problems such as expensive housing and pollution.

_____ **3.** Women in South Korea are not allowed to hold jobs outside the home.

_____ **4.** The most common religion in South Korea today is Christianity.

_____ **5.** Nearly 100 percent of the land in South Korea can be farmed.

_____ **6.** North Korea's Communist Party controls the government.

_____ **7.** North Korea is not as densely populated as South Korea.

_____ **8.** North Korea is well known for having many universities.

_____ **9.** Most of the arable land in North Korea is owned by the state and farmed by cooperatives.

_____ **10.** Because it has outdated technology, North Korea finds it difficult to produce high-quality goods.

Southeast Asia

CHAPTER 29

SECTION 1

Matching • *(10 points each)* Match each of the following terms or places with the correct description by writing the letter of the description in the space provided. Some descriptions will not be used.

_____ **1.** mainland

_____ **2.** archipelago

_____ **3.** Indochina Peninsula

_____ **4.** Philippines

_____ **5.** Myanmar

_____ **6.** Pacific Ring of Fire

_____ **7.** Mekong

_____ **8.** typhoon

_____ **9.** rice

_____ **10.** natural rubber

a. Country whose northern section contains Southeast Asia's highest mountains

b. Resource whose largest producers are Thailand, Malaysia, and Indonesia

c. Greatest river in Southeast Asia

d. Western half of New Guinea

e. A region's main landmass

f. Important crop in Southeast Asia

g. Archipelago that lies between the Southeast Asian mainland and new Guinea

h. Country to which the Mekong River flows from southeast China

i. One of two peninsulas that lie on the Asian mainland

j. Weather system that brings heavy rains and powerful winds to the island countries

k. Area that includes Borneo, Java, Sumatra, and New Guinea and where earthquakes and volcanic eruptions are common

l. Large group of islands

Southeast Asia

CHAPTER 29

Multiple Choice • *(10 points each)* For each of the following, write the letter of the *best* choice in the space provided.

_____ **1.** The Khmer Empire was based in Angkor in present-day
    a. French Indochina.
    b. Cambodia.
    c. Laos.
    d. Vietnam.

_____ **2.** Which Southeast Asian country was the only one to remain independent after the Spanish-American War?
    a. Siam
    b. Malaysia
    c. Myanmar
    d. Laos

_____ **3.** During World War II, much of Southeast Asia was occupied by
    a. Japan.
    b. China.
    c. Great Britain.
    d. the United States.

_____ **4.** Which country gave up its colonies of Vietnam, Laos, and Cambodia in 1954?
    a. United States
    b. Germany
    c. France
    d. Italy

_____ **5.** The Philippines now has a government
    a. run by the military.
    b. elected by the people.
    c. headed by a dictator.
    d. headed by a king and queen.

_____ **6.** In the 1960s the United States sent troops to defend South Vietnam against
    a. South Korea.
    b. North Korea.
    c. Japan.
    d. North Vietnam.

_____ **7.** In 1975 Indonesia invaded the former Portuguese colony of
    a. East Timor.
    b. North Vietnam.
    c. Thailand.
    d. Myanmar.

_____ **8.** A conflict began in 1978 when Vietnam invaded
    a. Korea.
    b. Myanmar.
    c. Thailand.
    d. Cambodia.

_____ **9.** What is the most common religion in mainland Southeast Asian countries?
    a. Judaism
    b. Hinduism
    c. Buddhism
    d. Roman Catholicism

_____ **10.** The largest Islamic country in the world is
    a. the Philippines.
    b. East Timor.
    c. Singapore.
    d. Indonesia.

Southeast Asia

SECTION 3

Fill in the Blank • *(10 points each)* For each of the following statements, fill in the blank with the appropriate word, phrase, or name.

1. The largest cities in Southeast Asia are located along major _____.

2. The mainland's largest city is Bangkok, the capital of _____.

3. Much of Bangkok is connected by *klongs,* or _____, which are used for transportation and for selling and shipping goods.

4. _____ (Rangoon) is Myanmar's capital and major seaport.

5. Ho Chi Minh City was once known as _____ and was the capital of South Vietnam.

6. The most important food and crop in Vietnam is _____.

7. The country of Laos has a(n) _____ government.

8. _____ is the most important part of the Cambodian economy.

9. Among the mainland countries, _____ has the strongest economy.

10. _____, a former British colony, gained independence in 1948 and was known as Burma until 1989.

Southeast Asia

True/False • *(10 points each)* Indicate whether each statement below is true or false by writing *T* or *F* in the space provided. If the statement is false, use the space below to explain why.

_____ **1.** Singapore is the largest of the island countries.

_____ **2.** Jakarta, the capital of Indonesia, is the region's largest city.

_____ **3.** Singapore is one of the most modern and cleanest cities in the world.

_____ **4.** The capital of the Philippines is Kuala Lumpur.

_____ **5.** Europeans once called Indonesia the Spice Islands because of its many valuable spices.

_____ **6.** Large areas of Indonesia's tropical rain forest are burned for farming.

_____ **7.** Most farmers in the Philippines are poor and own no land.

_____ **8.** Singapore is the least economically developed country in Southeast Asia.

_____ **9.** Malaysia is the world's leading producer of palm oil.

_____ **10.** A sultan is the supreme ruler of a Hindu country.

India

CHAPTER 30

Fill in the Blank • *(10 points each)* For each of the following statements, fill in the blank with the appropriate word, phrase, or name.

1. The _____ run along India's northern border.

2. India has three main landform regions: the Himalayas, the Gangetic Plain, and the _____.

3. India's towering mountain range was created when two _____ plates collided.

4. The edges of the Deccan plateau are defined by the _____ Ghats and _____ Ghats.

5. India's most important river, the _____, flows down from the Himalayas.

6. Hindus call the Ganges the "_____ River" and consider it sacred.

7. The Brahmaputra River starts in the Plateau of _____ and flows through the far northeastern corner of India.

8. Areas in the Himalayas have _____ climates with snow and glaciers.

9. India's _____ Desert, near the border with Pakistan, is hot and dry year-round.

10. Most of the people who live in India work in _____.

India

True/False • *(10 points each)* Indicate whether each statement below is true or false by writing *T* or *F* in the space provided. If the statement is false, use the space below to explain why.

_____ **1.** The first urban civilization on the Indian subcontinent was centered around the Indus River valley.

_____ **2.** The language of Sanskrit is still used today in India in religious ceremonies.

_____ **3.** The Muslim kingdom established in Delhi in the early 1200s was known as the Delhi empire.

_____ **4.** The founder of the Mughal Empire was Babur, whose name meant "the Lion."

_____ **5.** The Mughal Empire was reunited by Babur's grandson, Akbar.

_____ **6.** The famous Taj Mahal was built by Shah Jahan as a tomb for his beloved wife.

_____ **7.** During the 1700s and 1800s the French slowly took control of India.

_____ **8.** Sepoys were Indian troops commanded by British officers.

_____ **9.** Mohandas Gandhi used a strategy called nonviolent mass protest to seek independence for India.

_____ **10.** In 1947 the British divided their Indian colony into two independent countries—India and Pakistan.

CHAPTER 30 India

Matching • *(10 points each)* Match each of the following terms or places with the correct description by writing the letter of the description in the space provided. Some descriptions will not be used.

_____ **1.** Hinduism

_____ **2.** reincarnation

_____ **3.** karma

_____ **4.** Buddhism

_____ **5.** nirvana

_____ **6.** Jainism

_____ **7.** Sikhism

_____ **8.** castes

_____ **9.** Dalits

_____ **10.** green revolution

a. Belief that the soul is reborn again and again in different forms

b. Escape from the suffering of life

c. Groups of people who are born into their positions in society

d. Religion that teaches that all things in nature, including animals, plants, and stones, have souls

e. Mountainous region claimed by both India and Pakistan

f. Official national language of India

g. Positive or negative force caused by a person's actions

h. Program started by the Indian government to encourage farmers to adopt more modern methods

i. Religion founded in India by Siddhartha Gautama

j. People at the lowest level of the caste system

k. Religion that combines elements of Hinduism and Islam

l. Religion followed by about 80 percent of India's people

The Indian Perimeter

True/False • *(10 points each)* Indicate whether each statement below is true or false by writing *T* or *F* in the space provided. If the statement is false, use the space below to explain why.

_____ **1.** Bhutan, a tiny country high in the Himalayas, is north of Bangladesh.

_____ **2.** The Himalayas occupy about 10 percent of Nepal's land area.

_____ **3.** The world's highest mountain, Mount Everest, is located on Nepal's border with China.

_____ **4.** For centuries, invaders and traders traveled through the Khyber Pass.

_____ **5.** The Hindu Kush is Nepal's main farming area.

_____ **6.** The Indus Valley is Pakistan's main farming region.

_____ **7.** Bangladesh has one of the driest climates in the world.

_____ **8.** Storm surges are huge waves of water whipped up by fierce winds.

_____ **9.** Much of the country of Pakistan is desert and receives less than a foot of rain a year.

_____ **10.** Bangladesh's most important resource is oil.

The Indian Perimeter

Fill in the Blank • *(10 points each)* For each of the following statements, fill in the blank with the appropriate word, phrase, or name.

1. An ancient civilization developed in the _____ River valley around 2500 B.C.

2. Ever since Turkish invaders brought _____ to the area of Pakistan around A.D. 1000, it has continued to be the dominant religion.

3. In the early 1600s English merchants formed the English East India Company to increase

the _____ trade.

4. The capital city of Pakistan today is _____.

5. Marriages in Pakistan are usually arranged by the _____, and the bride's parents often pay money to the groom's parents.

6. Bangladesh is part of a region known as _____.

7. _____ became the independent country of Bangladesh in 1971.

8. Ninety-eight percent of the population in Bangladesh is made up of

_____.

9. The capital of Bangladesh, which is also its largest city, is _____.

10. _____ and disease are two of Bangladesh's largest challenges.

The Indian Perimeter

CHAPTER 31

SECTION 3

Multiple Choice • *(10 points each)* For each of the following, write the letter of the *best* choice in the space provided.

_____ **1.** Gautama, the Buddha, is believed to have been born in
  a. Nepal.
  b. Sri Lanka.
  c. Bhutan.
  d. the Maldives.

_____ **2.** Bhutan's most important resources are
  a. gold and silver.
  b. diamonds and other gemstones.
  c. oil and natural gas.
  d. timber and hydroelectricity.

_____ **3.** On what is the economy of Nepal based?
  a. shipbuilding
  b. agriculture
  c. high-technology
  d. oil refining

_____ **4.** The dominant religion in the Maldives today is
  a. Hinduism.
  b. Islam.
  c. Buddhism.
  d. Christianity.

_____ **5.** Most of the people in Bhutan are
  a. Christian.
  b. Jewish.
  c. Buddhist.
  d. Hindu.

_____ **6.** Today, about 90 percent of Nepal's population practices
  a. Buddhism.
  b. Confucianism.
  c. Daoism.
  d. Hinduism.

_____ **7.** Which group set up a trading post in Ceylon in the 1500s?
  a. Portuguese
  b. French
  c. British
  d. Spanish

_____ **8.** The main resources mined in Sri Lanka are
  a. bauxite and diamonds.
  b. granite and emeralds.
  c. graphite and precious gems.
  d. gold and silver.

_____ **9.** Nepal's capital and largest city is
  a. Bhutan.
  b. Kathmandu.
  c. Hindu Kush.
  d. Kyber Pass.

_____ **10.** What is the chief economic activity in the Maldives?
  a. farming
  b. mining
  c. tourism
  d. manufacturing

Australia and New Zealand

CHAPTER 32

SECTION 1

Fill in the Blank • *(10 points each)* **For each of the following statements, fill in the blank with the appropriate word, phrase, or name.**

1. _____ is the smallest, flattest, and lowest continent in the world.

2. Australia is the only _____ that is also a continent.

3. The _____ Range divides Australia's rivers into those that flow eastward and those that flow westward.

4. The Great Artesian Basin is Australia's largest source of underground

   _____.

5. The Great Barrier Reef is the world's largest _____ reef.

6. Dry _____ and steppe climates cover most of the country of Australia.

7. The kangaroos and koala that are native to Australia are _____, meaning that they carry their young in pouches.

8. The first humans to live in Australia were the _____.

9. Australia's Opera House is located in _____, Australia's largest city.

10. Australia supplies nearly half of the world's _____ used in clothing.

Australia and New Zealand

CHAPTER 32

SECTION 2

True/False • *(10 points each)* Indicate whether each statement below is true or false by writing *T* or *F* in the space provided. If the statement is false, use the space below to explain why.

_____ **1.** New Zealand lies southeast of Australia across the Red Sea.

_____ **2.** The highest mountains in New Zealand are the Southern Alps on South Island.

_____ **3.** New Zealand has a desert climate.

_____ **4.** Different kinds of bats and the kiwi are endemic species in New Zealand.

_____ **5.** New Zealand's first settlers came from Britain more than 1,000 years ago.

_____ **6.** New Zealand's government includes a prime minister and a parliament.

_____ **7.** The capital of New Zealand, Wellington, is located at the southern tip of North Island.

_____ **8.** Most New Zealanders speak English and are Christians.

_____ **9.** Most of New Zealand's industries and agriculture are located on South Island.

_____ **10.** New Zealand's largest city and seaport is Auckland.

The Pacific Islands and Antarctica

CHAPTER 33

Matching • *(10 points each)* Match each of the following terms or places with the correct description by writing the letter of the description in the space provided. Some descriptions will not be used.

_____ **1.** Tahiti

_____ **2.** New Guinea

_____ **3.** Greenland

_____ **4.** Irian Jaya

_____ **5.** Marshall Islands

_____ **6.** Antarctica

_____ **7.** ice shelf

_____ **8.** icebergs

_____ **9.** polar desert

_____ **10.** krill

a. Only island in the world that is larger than New Guinea

b. One of the most densely forested countries in the world

c. Huge chunks of ice

d. Name given to the eastern half of New Guinea

e. Example of an oceanic high island

f. About 98 percent of this continent is covered by ice

g. Shrimplike creatures eaten by the marine animals that live in Antarctica

h. Name given to the western half of New Guinea

i. High-latitude region that receives little precipitation

j. Example of a continental high island

k. Ledge that is formed over the water when ice reaches the coast

l. Area that includes two parallel chains of coral atolls

The Pacific Islands and Antarctica

CHAPTER 33

SECTION 2

Fill in the Blank • *(10 points each)* For each of the following statements, fill in the blank with the appropriate word, phrase, or name.

1. The large islands of _____ were the first Pacific Islands to be settled.

2. _____ was the first European to explore the Pacific region.

3. After _____ ended in 1918, Japan took over Germany's territories in the Pacific.

4. _____ are areas placed under the temporary control of another country until they can govern themselves.

5. U.S. territories in the Pacific include the Northern Mariana Islands, Guam, and

   _____.

6. Nearly two thirds all of Pacific Islanders live in _____.

7. Papua New Guinea's capital, _____, is Melanesia's largest city.

8. English or _____ is the official language on nearly all of the Pacific Islands, reflecting the region's colonial history.

9. _____ includes more than 2,000 tiny islands north of Melanesia.

10. The largest Pacific region is _____.

The Pacific Islands and Antarctica

SECTION 3

True/False • *(10 points each)* Indicate whether each statement below is true or false by writing *T* or *F* in the space provided. If the statement is false, use the space below to explain why.

_____ **1.** Antarctica is a continent that is also a country.

_____ **2.** In the 1770s British explorer James Cook sighted icebergs in the waters around Antarctica, suggesting the existence of a vast, icy continent.

_____ **3.** The first human expedition reached the South Pole in 1991.

_____ **4.** The "Antarctic Treaty" was designed to preserve the Antarctic "for science and peace."

_____ **5.** Researchers are the only people who live in Antarctica today.

_____ **6.** U.S. research stations in Antarctica include Palmer, on the Antarctic Peninsula, and McMurdo, on the Ross Ice Shelf.

_____ **7.** Studies have shown that carbon dioxide levels in the air have fallen over time.

_____ **8.** Scientists have found a thinning in the ozone layer above Antarctica.

_____ **9.** Antifreeze is a substance added to ice to keep it from turning into a liquid.

_____ **10.** A 1991 agreement forbids most activities in Antarctica that do not have a scientific purpose.

Answer Key

CHAPTER 1

Section 1 Quiz
1. F; People look at the world in different ways.
2. T
3. F; People familiar with geography are better able to understand the world around them.
4. T
5. T
6. F; They study issues at the local, regional, and global levels.
7. T
8. F; They cover larger areas.
9. T
10. T

Section 2 Quiz
1. c 2. g 3. j 4. a 5. f 6. h 7. i 8. e
9. b 10. k

Section 3 Quiz
1. human geography 2. Economic 3. physical geography 4. landforms 5. local
6. Cartography 7. computers 8. meteorology
9. climatology 10. climatologists

CHAPTER 2

Section 1 Quiz
1. c 2. l 3. e 4. i 5. j 6. g 7. a 8. k
9. f 10. d

Section 2 Quiz
1. Water 2. water 3. condensation
4. reservoirs 5. tributary 6. Groundwater
7. oceans 8. Pacific 9. floods 10. Dams

Section 3 Quiz
1. d 2. a 3. a 4. b 5. c 6. b 7. d 8. b
9. a 10. c

CHAPTER 3

Section 1 Quiz
1. sun 2. Tropic of Cancer 3. greenhouse effect 4. wind 5. air pressure 6. barometer
7. front 8. prevailing 9. currents 10. ocean

Section 2 Quiz
1. T
2. T
3. F; It is warm and rainy all year.
4. T
5. F; It receives little rain.
6. T
7. F; It provides an example of the humid subtropical climate.
8. F; They are found only in the Northern Hemisphere.
9. F; It is found in the tundra climate.
10. T

Section 3 Quiz
1. b 2. e 3. l 4. h 5. k 6. g 7. j 8. d
9. c 10. f

CHAPTER 4

Section 1 Quiz
1. T
2. T
3. F; They use them to add nutrients to the soil.
4. T
5. F; If too much salt builds up in the soil, crops cannot grow.
6. T
7. T
8. T
9. F; It is possible through planting or artificially seeding new trees.
10. F; It is occurring in the rain forests of Africa, Asia, and Central and South America.

Section 2 Quiz

1. Semiarid **2.** aqueducts **3.** aquifers **4.** desalinization **5.** pesticides **6.** oceans **7.** valleys **8.** acid rain **9.** ozone **10.** global warming

Section 3 Quiz

1. c **2.** h **3.** e **4.** a **5.** l **6.** f **7.** k **8.** b **9.** i **10.** g

Section 4 Quiz

1. d **2.** a **3.** b **4.** a **5.** c **6.** c **7.** d **8.** a **9.** b **10.** d

CHAPTER 5

Section 1 Quiz

1. Culture **2.** ethnic group **3.** multicultural **4.** Race **5.** culture traits **6.** acculturation **7.** environment **8.** domestication **9.** Subsistence agriculture **10.** civilization

Section 2 Quiz

1. c **2.** j **3.** h **4.** a **5.** l **6.** k **7.** g **8.** i **9.** f **10.** d

Section 3 Quiz

1. T
2. F; They differ from place to place.
3. T
4. F; It will hinder it.
5. F; They are not distributed evenly.
6. T
7. T
8. T
9. F; It is shrinking.
10. F; It is a nonrenewable resource and new supplies will eventually run out.

CHAPTER 6

Section 1 Quiz

1. contiguous **2.** Appalachians **3.** Ontario **4.** Continental Divide **5.** McKinley **6.** tropical

savanna **7.** steppe **8.** tundra **9.** Plains **10.** gas

Section 2 Quiz

1. d **2.** k **3.** f **4.** h **5.** l **6.** i **7.** a **8.** b **9.** g **10.** e

Section 3 Quiz

1. b **2.** d **3.** a **4.** b **5.** b **6.** d **7.** c **8.** c **9.** b **10.** b

CHAPTER 7

Section 1 Quiz

1. F; It is the world's second-largest country.
2. T
3. T
4. F; The St. Lawrence River links the Great Lakes to the Atlantic Ocean.
5. T
6. F; The mildest part of Canada is in the southwest.
7. F; Permafrost underlies about half of Canada.
8. T
9. T
10. F; Minerals are the most valuable of Canada's resources.

Section 2 Quiz

1. Vikings **2.** France **3.** British **4.** French **5.** French and Indian **6.** Provinces **7.** federation **8.** Métis **9.** Chinese **10.** Toronto

Section 3 Quiz

1. c **2.** h **3.** k **4.** f **5.** a **6.** l **7.** b **8.** i **9.** e **10.** j

CHAPTER 8

Section 1 Quiz

1. d **2.** d **3.** b **4.** d **5.** b **6.** a **7.** c **8.** c **9.** a **10.** c

Section 2 Quiz

1. T
2. T
3. F; Today they can read some Maya writing.
4. T
5. T
6. F; They defeated the Aztec.
7. F; They were called mestizos.
8. T
9. T
10. F; One major indicator is language.

Section 3 Quiz

1. congress 2. inflation 3. Agriculture
4. cash crops 5. 12 6. Mexico City
7. smog 8. oil 9. *maquiladoras* 10. slash-and-burn

CHAPTER 9

Section 1 Quiz

1. T
2. F; An archipelago is a large group of islands.
3. T
4. F; They are east of Florida.
5. T
6. T
7. T
8. F; They are common.
9. F; Volcanic ash enriches the soil and makes agriculture profitable.
10. T

Section 2 Quiz

1. Spain 2. mestizo 3. dictators
4. Guatemala 5. Belize 6. 15 7. civil war 8. Nicaragua 9. ecotourism
10. Panama

Section 3 Quiz

1. g 2. l 3. d 4. k 5. b 6. i 7. h 8. c
9. j 10. f

CHAPTER 10

Section 1 Quiz

1. Andes 2. cordillera 3. Guiana 4. *tepuís*
5. Llanos 6. Orinoco 7. piranhas 8. elevation 9. coffee 10. soil

Section 2 Quiz

1. T
2. T
3. F; The Chibcha were conquered by the Spanish.
4. F; It included Colombia, Ecuador, Panama, and Venezuela.
5. F; It became Colombia.
6. T
7. F; It is Bogotá.
8. F; It grows world-famous coffee.
9. T
10. T

Section 3 Quiz

1. e 2. j 3. b 4. l 5. d 6. h 7. a 8. k
9. i 10. c

Section 4 Quiz

1. b 2. d 3. a 4. b 5. c 6. d 7. a 8. a
9. b 10. c

CHAPTER 11

Section 1 Quiz

1. Argentina 2. Amazon 3. Chaco 4. Andes
5. Amazon 6. estuary 7. rain forest
8. Pampas 9. exhaustion 10. hydroelectric

Section 2 Quiz

1. a 2. d 3. d 4. b 5. c 6. a 7. b 8. c
9. c 10. d

Section 3 Quiz

1. f 2. i 3. k 4. b 5. d 6. g 7. l 8. h
9. c 10. j

Section 4 Quiz

1. F; The capital is Montevideo.
2. T
3. F; The main religion is Roman Catholicism.
4. F; More than 90 percent live in urban areas.
5. T
6. T
7. F; About 95 percent are mestizos.
8. T
9. T
10. F; Nearly half are farmers.

CHAPTER 12

Section 1 Quiz

1. j 2. c 3. l 4. a 5. k 6. g 7. e 8. h
9. b 10. i

Section 2 Quiz

1. F; Quinoa is a native Andean plant.
2. T
3. T
4. F; Their greatest achievement was the organization of their empire.
5. T
6. F; It was Chile.
7. T
8. T
9. T
10. F; It has had long violent periods in its history.

Section 3 Quiz

1. Ecuador 2. Quechua 3. La Paz
4. Indians 5. Peru 6. Machu Picchu
7. Lima 8. junta 9. Copper 10. NAFTA

CHAPTER 13

Section 1 Quiz

1. b 2. d 3. b 4. a 5. b 6. a 7. c 8. d
9. c 10. c

Section 2 Quiz

1. T
2. T
3. F; Athens was the first known democracy.
4. T
5. F; It was ruled from Constantinople.
6. T
7. F; It was occupied by Germany.
8. T
9. T
10. T

Section 3 Quiz

1. aqueducts 2. Rome 3. Latin
4. Christianity 5. pope 6. Renaissance
7. Rome 8. Galileo Galilei 9. Roman Catholic 10. grapes

Section 4 Quiz

1. b 2. h 3. i 4. l 5. a 6. j 7. g 8. d
9. c 10. k

CHAPTER 14

Section 1 Quiz

1. F; The Benelux countries are Belgium, the Netherlands, and Luxembourg.
2. T
3. T
4. F; It is also known as the Black Forest.
5. T
6. F; They have a highland climate.
7. T
8. T
9. T
10. F; The Alps provide hydroelectric power to those countries.

Section 2 Quiz

1. b 2. c 3. d 4. a 5. c 6. b 7. c 8. a
9. d 10. d

Section 3 Quiz

1. Franks **2.** Holy Roman **3.** Reformation
4. Prussia **5.** Adolf Hitler **6.** Holocaust
7. Berlin Wall **8.** Christmas **9.** Berlin
10. Ruhr

Section 4 Quiz

1. T
2. F; They won their freedom from Spanish rule.
3. F; Most were fought in Belgium.
4. T
5. F; They are mostly Roman Catholic.
6. F; It is Dutch.
7. T
8. F; It is famous for its tulips.
9. T
10. T

Section 5 Quiz

1. c **2.** h **3.** k **4.** f **5.** a **6.** i **7.** g **8.** b
9. l **10.** e

CHAPTER 15

Section 1 Quiz

1. F; Greenland is the world's largest island.
2. T
3. F; It is mostly covered by a thick ice cap.
4. T
5. T
6. T
7. T
8. T
9. F; Much of it has a marine west coast climate.
10. T

Section 2 Quiz

1. b **2.** a **3.** d **4.** b **5.** c **6.** a **7.** a **8.** d
9. a **10.** c

Section 3 Quiz

1. England **2.** Gaelic **3.** famine **4.** Republic
5. bog **6.** harp **7.** Roman Catholic
8. Patrick **9.** European Union **10.** Dublin

Section 4 Quiz

1. c **2.** g **3.** k **4.** d **5.** e **6.** a **7.** l **8.** j
9. h **10.** f

CHAPTER 16

Section 1 Quiz

1. T
2. T
3. T
4. F; Its most important river is the Danube.
5. T
6. F; It is called oil shale.
7. T
8. T
9. T
10. F; The region has serious pollution problems.

Section 2 Quiz

1. d **2.** i **3.** k **4.** a **5.** l **6.** b **7.** j **8.** h
9. f **10.** c

Section 3 Quiz

1. Ottoman Turks **2.** Roman Catholic
3. World War I **4.** Balkans **5.** Danube
6. agriculture **7.** Roma **8.** market
9. Bulgaria **10.** Albania

CHAPTER 17

Section 1 Quiz

1. F; It was the largest.
2. T
3. T
4. F; It is the Volga.
5. F; Nearly all of Russia is located at high
northern latitudes.
6. T
7. T
8. F; It has enormous resources.
9. T
10. T

Section 2 Quiz

1. d **2.** a **3.** c **4.** a **5.** b **6.** b **7.** d **8.** c
9. a **10.** d

Section 3 Quiz

1. European **2.** grains **3.** Moscow
4. Kremlin **5.** Light **6.** St. Petersburg
7. Volga **8.** German **9.** smelters
10. Urals

Section 4 Quiz

1. F; It is the largest.
2. T
3. T
4. F; It is sparsely populated.
5. T
6. T
7. F; Its most important industries are lumbering and mining.
8. T
9. F; It means "New Siberia."
10. F; It is known as the "Jewel of Siberia."

Section 5 Quiz

1. g **2.** c **3.** l **4.** a **5.** j **6.** e **7.** h **8.** d
9. k **10.** i

CHAPTER 18

Section 1 Quiz

1. F; They are part of a region called the Caucasus.
2. T
3. T
4. T
5. F; The Dnieper flows south through Belarus and Ukraine.
6. T
7. F; It has a humid continental climate.
8. T
9. T
10. F; They are found under the Caspian Sea.

Section 2 Quiz

1. f **2.** k **3.** a **4.** i **5.** c **6.** g **7.** j **8.** l
9. d **10.** b

Section 3 Quiz

1. Persian **2.** Soviet **3.** president **4.** Georgia
5. hydropower **6.** Turkey **7.** Agriculture
8. homogeneous **9.** agrarian **10.** fish eggs

CHAPTER 19

Section 1 Quiz

1. T
2. F; It lies north of some of the world's highest mountain ranges.
3. T
4. F; They are deserts.
5. T
6. T
7. F; It is the Syr Dar'ya.
8. F; It grew rapidly.
9. T
10. T

Section 2 Quiz

1. a **2.** c **3.** b **4.** b **5.** c **6.** d **7.** a **8.** b
9. d **10.** c

Section 3 Quiz

1. Russia **2.** New Year **3.** Kyrgyzstan
4. yurt **5.** Clan **6.** English **7.** Uzbekistan
8. citizenship **9.** civil war **10.** Persian

CHAPTER 20

Section 1 Quiz

1. g **2.** c **3.** j **4.** a **5.** e **6.** l **7.** h **8.** f
9. d **10.** i

Section 2 Quiz

1. c **2.** a **3.** d **4.** b **5.** c **6.** b **7.** a **8.** d
9. a **10.** b

Section 3 Quiz

1. Arabs   **2.** Ottoman   **3.** Saddam Hussein
4. Iran   **5.** Persian Gulf   **6.** oil   **7.** embargo
8. weapons   **9.** Tigris   **10.** Muslim

Section 4 Quiz

1. T
2. F; He took power under the title of shah.
3. F; They took American hostages.
4. T
5. F; Oil is its main industry.
6. T
7. T
8. T
9. T
10. F; The Taliban have forced women to stop working outside the home.

CHAPTER 21

Section 1 Quiz

1. Occupied Territories   **2.** Asia   **3.** Jordan
4. Dead Sea   **5.** Mediterranean   **6.** desert
7. Syrian   **8.** Negev   **9.** oil   **10.** mineral salts

Section 2 Quiz

1. c   **2.** g   **3.** l   **4.** i   **5.** a   **6.** j   **7.** d   **8.** k
9. h   **10.** b

Section 3 Quiz

1. F; It was established by the Hebrews.
2. F; It is known as the Diaspora.
3. T
4. T
5. F; It came under the control of Britain.
6. T
7. F; It is known as the Knesset.
8. T
9. F; The West Bank is the largest of the occupied areas.
10. T

Section 4 Quiz

1. b   **2.** c   **3.** d   **4.** a   **5.** a   **6.** a   **7.** d   **8.** c
9. d   **10.** b

CHAPTER 22

Section 1 Quiz

1. F; It stretches from the Atlantic Ocean to the Red Sea.
2. F; It comes from the Arabic word for "desert."
3. T
4. T
5. T
6. F; It helped the soil.
7. F; A desert climate covers most of the region.
8. F; Hardy plants and animals live there.
9. T
10. T

Section 2 Quiz

1. a   **2.** d   **3.** c   **4.** c   **5.** b   **6.** a   **7.** a   **8.** d
9. b   **10.** c

Section 3 Quiz

1. Egypt   **2.** fellahin   **3.** Cairo   **4.** Europe
5. Mediterranean   **6.** textiles   **7.** Suez
8. farmers   **9.** Overwatering   **10.** Islam

Section 4 Quiz

1. f   **2.** c   **3.** k   **4.** a   **5.** h   **6.** j   **7.** d   **8.** i
9. b   **10.** e

CHAPTER 23

Section 1 Quiz

1. zonal   **2.** Sahara   **3.** Sahel   **4.** harmattan
5. savanna   **6.** tsetse   **7.** coastal   **8.** Niger
9. bauxite   **10.** Oil

Section 2 Quiz

1. b   **2.** h   **3.** e   **4.** l   **5.** j   **6.** d   **7.** i   **8.** a
9. k   **10.** c

Section 3 Quiz

1. F; Most of them are Muslims.
2. T
3. T
4. T
5. F; Mali is landlocked.
6. F; Only about 3 percent of the land is good for farming.
7. T
8. F; They are among the poorest and least developed countries.
9. T
10. T

Section 4 Quiz

1. c **2.** c **3.** b **4.** d **5.** d **6.** a **7.** b **8.** a
9. d **10.** b

CHAPTER 24

Section 1 Quiz

1. F; It is a land of mostly plains and plateaus.
2. T
3. F; It is Africa's tallest mountain.
4. T
5. T
6. F; It is the source of the White Nile.
7. T
8. F; It is shallow.
9. T
10. T

Section 2 Quiz

1. g **2.** l **3.** i **4.** c **5.** e **6.** a **7.** k **8.** d
9. b **10.** j

Section 3 Quiz

1. d **2.** d **3.** c **4.** b **5.** a **6.** d **7.** a **8.** c
9. b **10.** b

Section 4 Quiz

1. 1980s **2.** Ethiopia **3.** droughts **4.** Muslim
5. Red **6.** Italy **7.** Somalia **8.** Islam
9. Djibouti **10.** Issa

CHAPTER 25

Section 1 Quiz

1. T
2. T
3. F; It flows westward to the Atlantic Ocean.
4. T
5. F; It is the Zambezi River that is famous for those things.
6. T
7. F; It is the trees that form a complete canopy.
8. T
9. T
10. F; The copper can be found in the copper belt.

Section 2 Quiz

1. b **2.** a **3.** b **4.** d **5.** c **6.** d **7.** a **8.** b
9. c **10.** b

Section 3 Quiz

1. Portuguese **2.** Belgium **3.** copper
4. 1960 **5.** Zaire **6.** civil war **7.** Kongo
8. French **9.** Kinshasa **10.** copper

Section 4 Quiz

1. l **2.** f **3.** h **4.** a **5.** d **6.** i **7.** b **8.** k
9. c **10.** g

CHAPTER 26

Section 1 Quiz

1. T
2. T
3. F; They are called the veld.
4. F; It protects many species.
5. F; Moisture is carried from the Indian Ocean.
6. T
7. F; They are deserts.
8. T
9. T
10. F; It is rich in mineral resources.

Section 2 Quiz

1. b **2.** g **3.** j **4.** e **5.** l **6.** c **7.** i **8.** k
9. h **10.** f

Section 3 Quiz

1. b **2.** a **3.** b **4.** c **5.** c **6.** d **7.** a **8.** b
9. d **10.** b

Section 4 Quiz

1. Namibia **2.** English **3.** Botswana
4. Tswana **5.** Salisbury **6.** whites
7. Mozambique **8.** Bantu **9.** France
10. subsistence

CHAPTER 27

Section 1 Quiz

1. c **2.** h **3.** j **4.** a **5.** l **6.** e **7.** i **8.** g
9. f **10.** d

Section 2 Quiz

1. d **2.** a **3.** d **4.** c **5.** b **6.** d **7.** a **8.** b
9. c **10.** c

Section 3 Quiz

1. China **2.** western **3.** 40 **4.** Beijing
5. Shanghai **6.** Hong Kong **7.** Portuguese
8. 10 **9.** iron **10.** most-favored-nation

Section 4 Quiz

1. T
2. F; It was the late 1200s.
3. T
4. F; It came under the influence of the
Soviet Union.
5. T
6. T
7. F; It was the search for spices that
brought them to Taiwan.
8. F; China took command of Taiwan.
9. T
10. T

CHAPTER 28

Section 1 Quiz

1. b **2.** c **3.** d **4.** d **5.** c **6.** a **7.** a **8.** b
9. b **10.** b

Section 2 Quiz

1. central Asia **2.** Shintoism **3.** Shamans
4. Confucianism **5.** samurai **6.** shogun
7. resources **8.** Korea **9.** Pearl Harbor,
Hawaii **10.** Diet

Section 3 Quiz

1. c **2.** f **3.** a **4.** l **5.** i **6.** d **7.** b **8.** g
9. j **10.** k

Section 4 Quiz

1. b **2.** b **3.** c **4.** a **5.** c **6.** c **7.** d **8.** a
9. d **10.** b

Section 5 Quiz

1. T
2. T
3. F; Women are beginning to hold important jobs outside the home.
4. T
5. F; Less than 20 percent can be farmed.
6. T
7. T
8. F; North Korea has only one university.
9. T
10. T

CHAPTER 29

Section 1 Quiz

1. e **2.** l **3.** i **4.** g **5.** a **6.** k **7.** c **8.** j
9. f **10.** b

Section 2 Quiz

1. b **2.** a **3.** a **4.** c **5.** b **6.** d **7.** a **8.** d
9. c **10.** d

Section 3 Quiz

1. rivers **2.** Thailand **3.** canals **4.** Yangon
5. Saigon **6.** rice **7.** communist
8. Agriculture **9.** Thailand **10.** Myanmar

Section 4 Quiz

1. F; Indonesia is the largest.
2. T
3. T
4. F; The capital is Manila.
5. T
6. T
7. T
8. F; It is the most economically developed country.
9. T
10. F; A sultan is the supreme ruler of a Muslim country.

CHAPTER 30

Section 1 Quiz

1. Himalayas **2.** Deccan **3.** tectonic
4. Eastern, Western **5.** Ganges **6.** Mother
7. Tibet **8.** highland **9.** Thar **10.** agriculture

Section 2 Quiz

1. T
2. T
3. F; It was known as the Delhi sultanate.
4. F; His name meant "the Tiger."
5. T
6. T
7. F; It was the British.
8. T
9. T
10. T

Section 3 Quiz

1. l **2.** a **3.** g **4.** i **5.** b **6.** d **7.** k **8.** c
9. j **10.** h

CHAPTER 31

Section 1 Quiz

1. T
2. F; They occupy about 80 percent.
3. T
4. T
5. F; Tarai is its main farming area.
6. T
7. F; It has one of the wettest climates.
8. T
9. T
10. F; Its most important resource is its fertile farmland.

Section 2 Quiz

1. Indus **2.** Islam **3.** spice **4.** Islamabad
5. parents **6.** Bengal **7.** East Pakistan
8. Bengalis **9.** Dhaka **10.** Flooding

Section 3 Quiz

1. a **2.** d **3.** b **4.** b **5.** c **6.** d **7.** a **8.** c
9. b **10.** c

CHAPTER 32

Section 1 Quiz

1. Australia **2.** country **3.** Great Dividing
4. water **5.** coral **6.** desert **7.** marsupials
8. Aborigines **9.** Sydney **10.** wool

Section 2 Quiz

1. F; It is across the Tasman Sea.
2. T
3. F; It has a marine west coast climate.
4. T
5. F; They came from other Pacific islands.
6. T
7. T
8. T
9. F; They are located on North Island.
10. T

CHAPTER 33

Section 1 Quiz

1. e   **2.** j   **3.** a   **4.** h   **5.** l   **6.** f   **7.** k   **8.** c
9. i   **10.** g

Section 2 Quiz

1. Melanesia   **2.** Ferdinand Magellan
3. World War I   **4.** trust territories   **5.** Wake
Island   **6.** Papua New Guinea   **7.** Port
Moresby   **8.** French   **9.** Micronesia
10. Polynesia

Section 3 Quiz

1. F; It is a continent only.

2. T

3. F; It reached it in 1911.

4. T

5. T

6. T

7. F; Carbon dioxide levels have risen over
time.

8. T

9. F; It is a substance added to liquid to
keep it from turning to ice.

10. T